Technologies of Mobility in the Americas

Intersections in Communications and Culture

Global Approaches and Transdisciplinary Perspectives

Cameron McCarthy and Angharad N. Valdivia
General Editors

Vol. 29

The Intersections in Communications and Culture series is part of the Peter Lang Media and Communication list. Every volume is peer reviewed and meets the highest quality standards for content and production.

PETER LANG
New York • Washington, D.C./Baltimore • Bern
Frankfurt • Berlin • Brussels • Vienna • Oxford

Technologies of Mobility in the Americas

EDITED BY
Phillip Vannini
Lucy Budd
Ole B. Jensen
Christian Fisker
Paola Jirón

PETER LANG
New York • Washington, D.C./Baltimore • Bern
Frankfurt • Berlin • Brussels • Vienna • Oxford

Library of Congress Cataloging-in-Publication Data
Technologies of mobility in the Americas / edited by Phillip Vannini ... [et al.].
p. cm. — (Intersections in communications and culture:
global approaches and transdisciplinary perspectives; v. 29)
Includes bibliographical references and index.
1. Migration, Internal—America. 2. Transportation—Social aspects—America.
3. Mobile communication systems—America. I. Vannini, Phillip.
HB1961.A3T43   303.48'3097—dc23   2012005685
ISBN 978-1-4331-1406-9 (hardcover)
ISBN 978-1-4331-1405-2 (paperback)
ISBN 978-1-4539-0776-4 (e-book)
ISSN 1528-610X

Bibliographic information published by **Die Deutsche Nationalbibliothek**.
Die Deutsche Nationalbibliothek lists this publication in the "Deutsche
Nationalbibliografie"; detailed bibliographic data is available
on the Internet at http://dnb.d-nb.de/.

The paper in this book meets the guidelines for permanence and durability
of the Committee on Production Guidelines for Book Longevity
of the Council of Library Resources.

© 2012 Peter Lang Publishing, Inc., New York
29 Broadway, 18th floor, New York, NY 10006
www.peterlang.com

All rights reserved.
Reprint or reproduction, even partially, in all forms such as microfilm,
xerography, microfiche, microcard, and offset strictly prohibited.

Printed in the United States of America

Table of Contents

Acknowledgments vii

Introduction

Chapter 1: Technologies of Mobility in the Americas: Introduction
Phillip Vannini, Lucy Budd, Ole B. Jensen, Christian Fisker, and Paola Jirón 1

Part I: Mobility, Technologies, and the Assemblage of Place

Chapter 2: Virtual Caribbeans: Edens, Economies, Elsewheres
Mimi Sheller 23

Chapter 3: Knowing Flows: How Migration
Research Meets Mobilities Through Digital Technology
Rob Shields 43

Chapter 4: If Only It Could Speak: Narrative Explorations
of Mobility and Place in Seattle
Ole B. Jensen 59

Chapter 5: How Car Drivers Took the Streets: Critical
Planning Moments of Automobility
Nick Scott 79

Chapter 6: Selling the World: Airline Advertisements
and the Promotion of International Aeromobility in
National Geographic, 1964-2004
Lucy Budd 99

Part II: Mobile Selves

Chapter 7: Solidarity on the Move: Technology, Mobility,
and Activism in a Hospitality Exchange Network
Jennie Germann Molz 119

Chapter 8: Glimpses of Motility of the Networked Self Across the Life Course
Christian E. Fisker 139

Chapter 9: Seniors, Cell Phones, and Tactical Restriction
Kim Sawchuk and Barbara Crow 157

Chapter 10: Haunting Technologies: Performing Memories of Place Through Effervescent Mobilities
Phillip Vannini and Rhys Evans 175

Part III: Technology, Technicians, and Mobility in Everyday Life

Chapter 11: Mobile Phones as a "Necessary Evil": Canadian Youth Talk About Negotiating the Politics of Mobility
Tamara Shepherd and Leslie Regan Shade 199

Chapter 12: A Sociology of Traffic: Driving, Cycling, Walking
Jim Conley 219

Chapter 13: Imaginative Technologies of (Im)mobility at the "End of the World"
Noel B. Salazar 237

Chapter 14: Technology and Technicians Out of Control: The Implementation of Transantiago From a Daily Mobility Point of View
Paola Jirón 255

About the Contributors 279

Index 285

Acknowledgments

This edited book would not have come together without the support of many friends, colleagues, and supporting partners. First and foremost we wish to thank the Social Sciences and Humanities Research Council (SSHRC) of Canada, which funded the "Cultures of Mobilities" conference held in Victoria, British Columbia in 2010, out of which this book arose.

We also want to thank Steven Bill for his work on the formatting of the book, Lindsay Vogan for her indexing, and Jonathan Taggart for providing us the cover images. Tony Budd gave the manuscript a thorough and thoughtful read and helped us edit and proofread.

Finally we are indebted to Peter Lang Publishing editor Mary Savigar for her trust and guidance and Design and Production Supervisor Sophie Appel for her patience and sharp eye.

1
Technologies of Mobility in the Americas: Introduction

Phillip Vannini, Lucy Budd, Ole B. Jensen, Christian Fisker, and Paola Jirón

The claims that *we* live in a mobile world, that *we* live mobile lifestyles, and that mobile technologies have changed the way *we* relate to place, time, and to one another should, by now, surprise no one. *Mobility*, indeed, has become a buzzword of the day. From TV commercials touting the capabilities of the latest electronic gadget to new ownership laws facilitating time-share agreements between similarly cosmopolitan, mobile leisure seekers, and from ubiquitous iPods and iPads to the latest startup low-cost airline, mobility in all its multiple shapes seems to rule the world. Take, for example, the very context in which the opening words of this chapter are being typed.

As I (Phillip Vannini) write the first few lines of this introduction on my laptop, I occasionally glance to the windows on my right, trying to catch a glimpse of the ferry boat that is supposed to arrive in a few minutes. I am headed to Philadelphia to present findings from my latest research study on mobilities and regional culture. There I will meet old and new friends and colleagues converging on the northeast USA from the four corners of the world. But to get there I will need to first cross the Salish Sea with our slow commuter ferry, find the number 620 city bus upon disembarking, get off that bus at the new Sky Train terminal, hop on the elevated train, exit at the airport stop, make my way through airport check-in and security, and finally find my seat on the plane. It will be a long flight, so I will have time to write a few more words and hopefully manage to get some shut-eye as well. Then it will be time to land, clear customs, grab a cab, and upon arrival at the university find a wireless signal to Skype home to let my family know I have arrived.

This might seem like an example of contemporary mobility colored by my position and social class as an academic. But mobility does not have to be so cosmopolitan, and it is perhaps in more ordinary and less cosmopolitan and glamorous manifestations that mobility technologies seem even more potent. Take the process of getting to the ferry terminal here on Vancouver Island—hardly the stuff of world travel. To get here I drove my car for about fifteen minutes, listening to my car radio and stereo en route. As I drove towards the outskirts of Nanaimo I pulled off the highway for some coffee. I drove through the Tim Hortons's drive-through, grabbed my coffee to go (obviously), and sipped it as I cruised to the Duke Point terminal. The fact that I was able to buy and consume my coffee this way reveals profound insights about the widespread character of *our* mobile technoculture. Not only was the Tim Hortons's coffee shop equipped for my unwillingness to get out of my car and walk through the doors—thus conveniently providing me with both a throwaway paper cup and a drive-through entrance, but the makers of my Japanese car had thoughtfully predicted that I would want to drink my coffee while on the move, and had therefore instructed the co-workers at their Indiana-based plant to build two convenient cupholders near the gear shift. Clearly, one does not have to be a member of the international jet set to call oneself a citizen of a mobile world. Indeed, mobility is very much a part of life as we know it.

Yet despite its taken-for-granted character, its seemingly ubiquitous presence, and the habitual reliance that *we* place on cupholders, car stereos, transit buses, Skype, and interpersonal networks of friends and co-workers, mobility is not something to be nonchalant about. Indeed *we* are differentially mobile. The gentleman sitting next to me in the departure hall, talking on his mobile phone, is telling the person on the other end of the line to grab his or her passport for a quick, extemporaneous getaway to Seattle. The older ladies just in front of me have different concerns: they are frantically looking to borrow a wheelchair from the ferry terminal staff in order to compensate for the lack of elevator access to the boat. In the meantime a group of foreign exchange college students have arrived. Their world Englishes, spoken with varying accents and different levels of confidence, tell stories of different walks of life, but arguably share a common orientation to being worldly and mobile. As their voices become louder in the increasingly crowded departure hall, two teenage girls momentarily halt their frenetic texting to see what the raucousness is about. As this picture of ordinary mobile life unfolding around me becomes more vivid through my description, and more complex, who *we* are and what *our* mobilities mean begin to blur.

Technologies of Mobility: The Basics

There are many of *us*—indeed so many that our collective identity has been put in scare quotes. The diversity of the citizens of this techno-mobile world cannot escape our attention. In the short vignette described above there are two teenagers, a group of ESL students, a businessman, two elderly ladies, and me. There are also people in the background, including those who produced and sold me my coffee, built my car, sold me my ferry tickets, and received my many phone calls and text messages. All these individuals are what we might call *technicians*. They are not technicians in the ordinary sense of the word, but rather technicians in a more basic denotation of the term: people with the skills to operate tools intended for the satisfaction of particular purposes. Drivers are technicians, so are cell phone operators, laptop computer users, ferry passengers, wheelchair users, and many of the other individuals introduced directly or indirectly in this vignette. There are many different types of technicians, of course, but for the most part their roles always imply a common form of engagement with their tools, an embodied engagement we might call *use*.

The tools which technicians use can be referred to as *technics*. Technics are tools, implements, technologies, objects, instruments, or otherwise anything—material in a more or less obvious way—that enable technicians to carry out their actions. A ship's whistle—which is sounding as I am typing these words—is a technic that allows boaters nearby to be alert to the moving presence of this vessel. A technic's purpose differs, of course, depending on the orientation, need, and purpose of the actions of the technician involved. For a sail boat operator, the ship's whistle is a potential lifesaver. For the captain and crew it is more like a tool to prevent a lawsuit, or perhaps, more kindly, a tool to demonstrate thoughtfulness and respect for the smaller guys out on the water. For me right now it is both a momentary distraction and a good example. Technics, therefore, are just as variable in their identity and role as are technicians. Their "being" is relational, and can only be understood by examining the ecology of the relationships in which they are enmeshed.

Technicians use technics in different ways. These different ways are a matter of purpose and functionality of the technic, but also a matter of *technique*. Techniques are the ways, the styles, the procedures, the processes, the systems (and much more) in which particular technicians use their technics. At times techniques are dependent on technicians' *know-how* and/or on the technical *affordances* of their tools. For example, my laptop computer is currently being put to use in a way that resembles that of a glorified typewriter. The young woman sitting in the seat in front of me is instead using her laptop to draw what seems like some kind of architectural blueprint for a house. I would not know where to begin to use my computer in that way. But, beside know-how, technique can also be at other times a manifestation of style, and therefore of the identity, personality, and comportment of the user. For instance the man sitting behind me has been using his cell phone to loudly scold someone who seems to be his son. Whereas—in a different communication style—the much more reserved and quiet, older gentleman seated almost immediately to my right has quickly picked up his ringing cell phone, excusing himself to the caller (and, implicitly, to the rest of us in the lounge), as he walks away to find a more private spot where he can talk.

Technics, technicians, and techniques form the three basic cornerstones of what this book is about. Whether it is cell phones or airplanes, passengers or magazine advertisements, what the chapters collected here have in common is a basic orientation towards technology as a set of relations. This *relational* approach to technology demonstrates that technology is neither a utopian nor a dystopian force driving the universe towards progress or involution. For all the contributors to this book, what mobile communication technologies like smart phones and spatial mobility technologies like cars and ferry boats have in common is the *potential to transform*—and be transformed by—the very relationships in which technics and technicians are involved. Nothing more, and nothing less. Technology, therefore, becomes in this context a substitute term for a *sociotechnical assemblage* in which multiple components play different roles based on circumstances, context, purposes, needs, affordances, material possibilities, and multiple other contingencies and variables.

Much has been written, of course, on the theoretical and even philosophical foundations for conceiving technology as being relational. As of late, a great deal of knowledge has also begun to accumulate on the specifically mobile elements of these relations. This book draws upon these bodies of knowledge and intends to contribute to them, but with a particular academic objective in mind. The study of mobilities writ large has recently begun to examine the social, political, historical, cultural, economic, geographic, communicative, and material dimensions of movement in which the contributors to this book are interested. Concentrating on the intersecting movements of bodies, objects, capital, and signs across "time-space" (Thrift, 2008), and dissecting how practices, experiences, representations, and political dynamics shape new networks and lifeworlds, students of these multiple forms of mobility have examined the simultaneously technological and cultural (in one word: technocultural) processes underpinning many different forms of mobility. But for the most part these students of mobility have concentrated their research efforts on the northern European—and particularly British and Scandinavian—social context.

Whereas in Europe the study of mobilities has begun to take a strong hold in academic units, professional research networks, and recognized publication outlets, the study of mobilities is still in its adolescence in the Americas. To answer the need for knowledge sensitive to the unique characteristics of the North, Central, and South American context the Pan-American Mobility Network was founded in 2009 by a group of scholars and students working across the social sciences and humanities—especially communication, cultural studies, human geography, sociology, and cultural anthropology. That network held its first meeting in Victoria, British Columbia, in the spring of 2010. Born from that conference, and the collectively perceived need to offer the first book discussing mobility and communication technologies in the context of the Americas, are the writings included herein. As the same network of scholars—at the time of writing—

converges in Philadelphia to cement the ties fostered in Victoria and develop stronger linkages, the field of mobilities research appears more vibrant than ever before.

While its boundaries are to a great extent still the subject of debate (see Adey, 2006, 2009; Canzler, Kaufmann, & Kesselring, 2008; Kaufmann, 2002; Kellerman, 2006; Larsen, Urry, & Axhausen, 2006; Urry, 2007), many would agree that the study of mobilities rightly encompasses sociotechnical processes that focus on the material, imaginative and/or virtual movement of people, signs, and objects (see Hannam, Sheller, & Urry, 2006; Sheller & Urry, 2006; Urry, 2000, 2007). This means that mobility studies can focus on phenomena as diverse as migration; transport; travel and tourism; the social organization and experience of transportation and communication; smart phone use; regional and transnational flows of capital and material objects (e.g., raw materials, manufactured resources and consumer goods); infrastructure ecologies; and much more. Aside from the contemporary relevance of such topics—indeed even their "fashionableness," some critics might opine—the study of mobilities has prompted crucial, key developments in communication and cultural studies and in sociological, anthropological, and geographical debate in the light of the challenges it poses to key concepts including place, time, society, and culture, with their older connotations of structures as being frozen and static. For this reason many influential academics now refer to a "mobilities turn" or a "mobility paradigm" (Hannam et al., 2006; Sheller & Urry, 2006; Urry, 2000, 2007).

For Mimi Sheller—one of the contributors to this volume—and John Urry (2007) mobility studies directly challenge the social scientific agenda as a whole. For them the study of mobilities constitutes a new paradigm because it prompts us all to question the sociocultural significance of movement for both theoretical and methodological development. The world is increasingly on the move, they reminded us:

> asylum seekers, international students, terrorists, members of diasporas, holidaymakers, business people, sports stars, refugees, backpackers, commuters, the early retired, young mobile professionals, prostitutes, armed forces; these and many others fill the world's airports, buses, ships, and trains. The scale of this traveling is immense. (Sheller & Urry, 2006, p. 207)

This new technoculture alerts us to the existence of new ways of interacting; ways which are rapidly providing the world's population with new challenges and new opportunities (which, in so doing, may also create new conditions of inequality).

> As Sheller and Urry (2006) asserted, the new mobilities turn challenges the ways in which much social science research has been 'a-mobile' . . . [and] failed to examine how the spatialities of social life presuppose (and frequently involve conflict over) both the actual and the imagined movement of people from place to place, person to person, event to event. (p. 208)

Similarly, a new mobility paradigm challenges the way we understand the sociotechnical assemblages that make up *our* many technocultures of the present time. While avoiding the excesses of a nomadic outlook that posits life as exaggeratedly placeless and traditionless (Cresswell, 2001, 2006), and while recognizing what is old about this "new" perspective, the mobility turn in which contributors to this book situate themselves emphasizes the importance of flows, networks, connections, immobilities, movements, performances, processes of deterritorialization and reterritorialization, transnational organizations, immobile infrastructures, and even fixed moorings. Focusing on case studies of the relations between technics, technicians, and techniques in the context of the Americas at the levels of the individual, place, and everyday life, these reflections prompt us to consider the sociotechnical assemblages through which people choose, experience, practice, perceive, negotiate, engage, adapt to, cope with, and struggle over the meanings of mobility and communication.

Mobility and Technology in the Americas: A Brief Historical Background

Technology, in all its diverse manifestations, has facilitated new ways of becoming and being mobile. It has transformed how, when, and where we work; where we live; how we spend our leisure and vacation time; what we eat; the commodities that we purchase; and the material and cultural products that we consume. While the fusion of technology and mobility dates back to the earliest days of human existence, it is only within the last 150 years or so that the application of mobility technology has accelerated the compression of time and space. Originally, mobility was a prerequisite for survival. Our ancestors had to move to find food and water, seek shelter, flee from enemies, escape from natural disasters, and colonize new territories. Over time, horses and dogs were domesticated and used to transport people and goods over ever greater distances. Trees were felled and fashioned into rafts and canoes to cross rivers and estuaries. At length, ocean-going sailing vessels were developed and pioneering explorers, settlers, and merchants, embarked on long-distance voyages to far-off lands in the hope of establishing new settlements and fostering new trading links. Later centuries saw the invention of powered transport and the development of road, rail, and, air networks. Today, people move not only because they need to, but also simply because they can. Volumes of discretionary and leisure travel among citizens of more affluent "Western" nations grew rapidly throughout the twentieth century to the point where a fortnight's vacation on a tropical beach has become enshrined as something of a "right."

Academics have traditionally tackled the subject of mobility by focusing on the technology and infrastructure that supports it but, following the mobilities turn in the social sciences, scholars began to systematically examine the unique spaces and socialities that different forms of transport and mobility engender. These studies, which originated from different disciplinary backgrounds and employed a wide variety of theoretical and methodological techniques, have variously explored the corporeal experiences of flying, cycling, driving, mobile sightseeing, walking-and-phoning, commuting, emigrating, and voyaging by rail, road, and sea. Their existence demonstrates the extent to which academia is becoming increasingly attuned to the social significance of different forms of movement.

The development of mechanized and vehicular transport during the nineteenth and twentieth centuries undoubtedly represented a crucial stage in the development of mobility cultures in the Americas. Steamships, railroads, private automobiles, and, later, commercial aircraft transformed the physical landscape and enabled growing numbers of people to travel to more places more quickly, more cheaply, and more easily than ever before. Mobility was promoted as a modern and productive social and economic force that would enable people to do more things more quickly and more efficiently than ever before while simultaneously permitting exposure to experiences and opportunities that would have been unimaginable to previous generations. Reflecting the spirit of modernity, progress, and optimism, vast railroads, harbors, multilane highways, parking lots, and airport terminals were constructed. Yet, in addition to transforming the appearance of the physical landscape, the presence of these mobility infrastructures impacted on the social fabric and material culture of everyday lives in subtle and complex ways that academics have not always been quick to highlight. As Jensen (this volume) argues in his study of the Sea Wall, an elevated section of State Route 99 that cleaves through downtown Seattle, individuals who live adjacent to such infrastructures often exhibit very complex and often contradictory emotional attachments towards them. Consequently, while ships, trains, cars, and planes metaphorically "shrink" time and "annihilate" space for their occupants, the fixed infrastructures of these spaces of flow undoubtedly affect the lives of those living alongside them, whether through their visual presence, or through combinations of noise, air pollution, safety fears, or other concerns their use generates.

In addition to impacting on the lives of those on the receiving end of these technologies, routine international and intercontinental mobility has enabled the large scale movement of people both within and beyond the territorial confines of American nations—thus changing the very nature of places. In the United States and Canada, the creation of transnational railroad systems was instrumental in linking the nations and fostering a sense of national unity. These railroads not only enabled fresh produce and raw materials to be transported from their points of origin or manufacture to distant markets and "opened up" formerly inaccessible regions to vacationers and business travellers, they also showcased technological achievement and human endeavour. Tunnels, stations, embankments, cuttings, and bridges became often-celebrated physical manifestations of human ingenuity and the apparent "conquest" of nature. In the Canadian province of British Columbia, for example, public information boards recount the construction of the famous "spiral tunnels" and commemorate the symbolic driving of the "last spike" at Craigellachie which marked the completion of the Canadian Pacific Railway across the Rockies.

Perhaps as a result of their ubiquity and familiarity, very few infrastructures of automobility have achieved similar cultural status. This is, perhaps, unfortunate, as arguably no other ground transport mode did more to shape the Americas than the invention of the automobile. Famously associated with the work of American entrepreneur, Henry Ford, and his iconic Model T, the automobile came to embody all that was exciting and progressive about American society. The automobile quickly became a central feature of American popular culture and was eulogized in song, movies, novels, and verse. That the development and widespread utilization of the private automobile fundamentally changed the landscape of the Americas is not disputed, as new freeways, parking lots, motels, and drive-through, roadside diners sprang up to serve the perceived needs of a new and increasingly mobile community. What, until recently, has been less certain is how the automobile changed human experiences of movement.

Thanks to the seminal work of John Urry and Mimi Sheller the automobile is not only conceptualized in terms of what it does but also in terms of what it means to those who use it (see also the chapters by Scott and Conley in this volume, for example). This new body of literature about cultures of automobility has examined how the private car is socially constructed, utilized, and experienced by those who use it and those who are affected by its presence, whether as drivers, passengers, other road users, or pedestrians. This awareness of the multiplicity of automobile experiences has recently been applied to other transport modes, leading to useful considerations of the extent to which other technologies of mobility, including passenger ferries, trains, and aircraft affect the embodied and emotional states of users.

One of the most profound new experiences of mobility is flight. Although the Wright brothers had demonstrated that human-powered flight was possible as early as 1903, it was not until the 1920s that regular commercial services began in earnest. However, the limited speed and range of the early aircraft, combined with the expense of this new form of transport, meant that aviation was available only to a wealthy few and it was not until after World War Two, when innovations in propulsion, aerodynamics, and material sciences significantly lowered the financial cost of flight, that the era of mass air travel began. From a few thousand passengers in 1945, the global airline industry has grown into a multibillion dollar enterprise that facilitates the routine international mobility of over 2.4 billion passengers and millions of tons of airfreight every year (for the cultural significance of this growth see Budd, this volume).

Despite growing competition from airports in China and the Far East, North America remains the undisputed hub of global commercial aviation. Thirteen of the 30 busiest airports in the world are located in the United States and the vast interlocking web of air routes that radiate out from the country has been responsible for broadening travel horizons, assisting the rapid worldwide diffusion of U.S. culture and enabling growing numbers of American citizens to live a semi-nomadic lifestyle predicated on the ready availability of air travel.

The history of mobility has been one of constant development and specialization as machines were devised to overcome particular topographic or physical challenges in an effort to bring people and places ever closer together in time and space. Each sovereign nation within the Americas has developed a transport and mobility system to suit its own needs and exhibits very diverse manifestations of mobility. However, in addition to facilitating the mobility of people, goods, and information around the world, transport technologies also enable the transnational mobility of smugglers, migrants, infectious diseases, narcotics, criminals, and terrorists. Furthermore, as numerous fatal air, rail, marine, and road crashes have shown, technology and the people who design and operate it are not infallible. When technology fails, society's foundations are (albeit temporarily) shaken—a key testament to its social and political importance.

Mindful of the historical and regional contexts in which different mobilities have unfolded within the Americas, and conscious of the various social and material relations that old and new technologies have shaped and reshaped—and, in turn, of how the technologies themselves have been shaped by these relations—contributors to this volume examine processes pertaining to three aspects of mobility technologies in particular: place, selfhood, and everyday ways of life. The book is divided into three distinct, yet intrinsically interrelated, sections that correspond to these core foci. Some chapters are predominately empirical in nature while others are more theoretical, yet all point to the need to better understand technologies of mobility as relational and contextual.

Mobility, Technologies, and the Assemblage of Place

Chapters in the first section of this book focus on the complex relationships between technologies, mobilities, and the way we understand places. Increasingly, the literature on mobilities has begun to pay attention to the ways in which networks and infrastructures create new cultural manifestations and practices that not only reconfigure how we relate to other social groups and cultures, but equally importantly, how this process (re)shapes our understanding of place. Since the very early attempts to theorize place and related dynamics we find a division between a perspective focusing on place as static and fixed, and places understood as mobile and in flux. In other words, there is a distinction between a sedentary and a nomadic understanding of place. Thus, for example, in their respective chapters Sheller looks at the Caribbean archipelago as a moving, evolving place, and Shields examines how these understandings have informed our comprehension of migration as a form of mobility.

However, thinking beyond the dichotomies of the fixed and the fluid we may begin to orient our analytical gaze towards the ways in which mobility technologies, in all their diverse manifestations, transgress the division between culture and nature, and technology and society. Drawing upon theoretical insights from areas such as Actor-Network Theory (Latour, 2005), Non-Representational Theory (Thrift, 2008), and Assemblage Theory (DeLanda, 2006), a new agenda for mobilities may be arising (see, for example, Farias & Bender, 2010). En route to this renewed interest in how technologies and materialities are shaped by mobilities of various sorts (from global communication via freight corridors to the everyday commute), we may be witnesses to a perspective that understands infrastructures as complex networks of artifacts that assemble human as well as non-human entities, ultimately challenging our understanding of place. Such complexity is obvious, for example, in Jensen's analysis of Seattle's Sea Wall—through which the very character of human and non-human blurs—as well as in Scott's chapter, in which historically changing planning paradigms contribute to the evolution of urban space.

The literature on these issues has for a while been firmly located within human geography and sociology (e.g., Cresswell, 2006; Urry, 2007). However, new insights from different disciplines that may start to emerge under the mobility turn point towards the inclusion of the arts, architecture, and the built environment. Drawing upon a recent publication from the architectural company, Infranet Lab / Lateral Office, Keller Easterling (2011) pointed towards the potential for conceiving infrastructure as comprising networks which transcend geometry and protocol. Easterling stated: "While infrastructure typically conjures associations with physical networks for transportation, communication, or utilities, it also includes the countless shared protocols that format everything from technical objects to management styles to the spaces of urbanism (Easterling, 2011, p. 10).

The *protocol* dimension of infrastructure points towards an understanding of mobile technologies and their networks as being highly influential *software*—a point illustrated in Sheller's chapter, for example. Or, put differently, this is a dimension concerning techniques in the sense of manuals, protocols, and action guidance for how to use the hardware and objects in question. Equally importantly, the *geometry* dimension points towards the fact that these systems are material and physical manifestations of *hardware*. This dimension speaks to the technics of infrastructures understood as material artifacts and physical objects.

To understand the meaning of mobility within these contemporary networks and infrastructures is to comprehend the new and complex relationships that exist between entities that were previously thought to belong to separate realms; culture/nature, city/countryside, human/non-human, and technology/society—amongst others. What is taking place within these complex network architectures of mobilities and technocultures is that basic categories are being reconfigured. One such category is that of *place*. Places are now best understood as assemblages of technologies, cultures, and mobilities and they need to be theorized like this in order for the analysis to progress in accordance with the increasing complexity of contemporary society (Dovey, 2010). As tools and artifacts are being negotiated by human and non-human agents in accordance with protocols and specifications that orchestrate the movements of people, goods, vehicles, and signs, we find ourselves looking at technological assemblages. Needless to say, such complex phenomena may influence not only how we relate to our environment but also to other people. More profoundly, however, we would claim that these dynamics also have repercussions for our understanding of place in a material sense.

These developments may be seen all over the globe and indeed they are the hallmark of what we, in the absence of a better term, may call *globalization*. However, arguably nowhere else is the diversity and complexity of the entanglement of mobility and technoculture more dynamic than in the Americas. So, in order to move beyond a certain European hegemony in the mobilities literature (see Jensen, 2009), and partly to pay homage to the multiplicities and dynamics of this region, this volume brings together theories and studies from a wide array of disciplines (and an equally diverse number of settings) to show how the contemporary complexity of communication, mobilities, and technologies changes and challenges one of the key dimensions of society: place. Many of the chapters in this collection of mobilities research are thus indicative of a new way of perceiving the notion of place.

Take, for example, Budd's analysis of the changing discourses surrounding aeromobility and international travel. Her work shows how the way we relate to places—with our cultural bearings and normative understandings of what places are for, what types of actions and activities they will afford, as well as what forms of action and interaction will be excluded—becomes altered by this way of thinking about place. If place is an assemblage of various communication processes, and mobilities are a cultural and social set of practices being mediated by networks of technologies, systems of infrastructures and flows of objects, subjects and communication, then we would argue that the study of mobilities and technologies is as much a study of place and the meaning of place as it is a study of the ephemeral, the fluids, and the flows. Rather than thinking about place as static and sedentary the studies in this volume, and especially those in part one, point towards a different perspective of place: one which is networked, dynamic, and contingent upon a multitude of technologies, subjects, and cultures.

Mobile Selves

Mobility-enabling technologies and their related infrastructures can be seen as bringing about changes in human connections at multiple scales and vantage points. One of these many potential vantage points is from the perspective of the individual. At any given moment in time we occupy a situated individual body (replete with unique skills and relationships of various sorts with others where we may rely on their skills and possessions) which becomes enmeshed with technologies and infrastructures that traverse built environments and enable both physical and virtual connections. Yet, these individual elements are not static. They change and evolve, each in its own unique way and at its own pace. As Germann Molz's study of a community of couch-surfers examines, at any given moment in time and situated location, we may find ourselves attempting mobility through networks which may or may not encourage, or easily facilitate, the unique cocktail of body, identity, skills, possessions, assistance from others, technologies, infrastructures and built environment characteristics that we attempt to bring together in some form of assemblage, in order to attempt mobility. We may find ourselves in forms of stratification and segregated mobility patterns (Jensen, 2007) where we are able to pass relatively easily, in an almost frictionless sense, through the networks we traverse, while others may find themselves held back from entering a particular network, or struggling partway through an attempted network connection, or brought to a full stop somewhere within a particular network connection, owing to different assemblage configurations.

One way of bringing forward the differences in the self and its relationship with others is through a life course perspective. Hodge (2008) reminded us that the life course of each individual is unique, while at the same time it is affected by social forces and stratification, which can lead to some common life stages among the elderly, such as reduced work, followed by retirement, frailty and eventual death. Yet, the life course can be seen not simply as one particular person's experience as they travel across time. Utilizing an approach that is a core principle to a large group of life course researchers in North America, we can see that life is not lived in isolation, that it is experienced interdependently, creating linked lives, as Fisker's chapter explains.

We each begin the life course figuratively and skill wise, *naked*. As independent individuals, before we can draw on the skills and support of others (such as family and friends), and progressively acquire the skills that open up networks of mobility technologies and mobility infrastructures, we have *naked capacities* that keep our *reach* very close at hand (Ihde, 1990, p. 75). Our needs have to be within a very short reach of our hands and mouths or else we have to rely on our needs being delivered to us by others. As Sawchuk and Crow's chapter analyses demonstrate in the context of seniors' use of mobile phones, over time we build up complex connections of needs through ever-evolving, ever-adapting assemblages of body, skills, others, technologies, and infrastructures. Considered collectively, such assemblages can be seen as extensions of the self.

Over the last 70 years scholars from a variety of disciplines, including planning, architecture, media, anthropology, and philosophy of technology, have probed this notion of extension. In the mid-1930s, Mumford (1934) viewed tools and utensils as being extensions of the human organism. Merleau-Ponty (1962, p. xi) argued that "there is no inner man, man is in the world, and only in the world does he know himself." Ihde (2002, p. xi) saw the many complex, embodiment relationships that humans now exist within and concluded that one's reach has been extended well beyond one's naked capacities. Hall (1966/1990, p. 188) saw humans and their extensions as one interrelated system and opined that certain extensions may not be well suited to all and that we should pay more attention to the extensions we create and to whom they are well or ill suited. As Mitchell explained, we find ourselves extended and co-existing with others, across networks:

> Embedded within a vast structure of nested boundaries and ramifying networks, my muscular and skeletal, physiological, and nervous systems have been artificially augmented and expanded. My reach extends indefinitely and interacts with the similarly extended reaches of others to produce a global system of transfer, actuation, sensing, and control. My biological body meshes with the city; the city itself has become not

only the domain of my networked cognitive system, but also—and crucially—the spatial and material embodiment of that system. (Mitchell, 2004, p. 19)

We can open up these complex connections ever so slightly here by looking at some of the changing elements—the changing body, changing skills, changing relationships to others and the related extensions of the self that open, grow, adapt and potentially recede across the life course.

Our bodies change across our life course. We grow taller, with longer legs and arms. We go from sitting in baby seats to sitting in high chairs (getting into both with the assistance of others), and eventually we graduate to the world of adult scale chairs. What used to seem to our little bodies like big and heavy doors, which needed to be opened for us by others, turn into everyday objects that we pass through with ease. Still later in life, we may become frail and suffer from arthritis, restricted vision, reduced endurance, and failing balance. Doors that seemed easy to pass through in midlife once again seem heavy. It may be easy sit down on a chair, whether in a house, a car, or a train, yet difficult to get out of it.

At a very early stage in our life course we cannot even crawl. We are highly dependent on others for our mobility. Slowly we build up skills that enable crawling and walking. Gehl (2010) noted that life begins in earnest when we reach the ability to walk. In time, we reach an age where we have an opportunity to be tested on the skills required in order to hold a driver's license. Later in life, as the physical body degrades, walking may become problematic. This often taken-for-granted basic form of mobility, which is often wrapped inside larger mobility assemblages, can potentially fall into jeopardy, introducing new challenges to mobility. At the same time, the skills which used to be in place to open up a world of automobility may become constrained through self-regulation or through formal channels that result in the revocation of a driver's license. Similar to a child who has to be driven to school and social events, relying on the driving skills of a parent, friend, or neighbor, older adults may have to adapt connections and once again (albeit reluctantly and perhaps unwillingly) come to rely on the driving skills of others.

During the journey through life we may choose to avail ourselves of, or adorn our bodies with, technologies that assist mobility. Spectacles, contact lenses, and laser eye surgery can enhance poor or degrading vision. Hearing aids and cochlear implants amplify sound waves and can help overcome hearing difficulties. Knee, hip, and knuckle replacement surgery can restore pain-free mobility to damaged joints, and the introduction of a walking stick or a walking frame can assist with balance. Similarly, electric carts, mobility scooters, and/or wheelchairs, can be used to overcome limited mobility and fatigue, and enable users to maintain a degree of autonomy and independence. As Vannini and Evans show in their chapter, technologies of mobility accompany us from childhood to old age, weaving memories and binding ties. Still other technologies, when used together with mobility infrastructures, extend us well beyond our naked capacities. Wrapping the naked self with technology, Ihde (2002, p. 138) stated: "we are bodies—but in that very basic notion one also discovers that our bodies have an amazing plasticity and polymorphism that is often brought out precisely in our relations with technologies." Extending ourselves via skills, the assistance of others, technologies, and infrastructures, we can see the many complex forms of mobilities that enable, sustain, configure, and reconfigure all the connections described in this second part of the book.

Technology, Technicians and Mobility in Everyday Life

Part three of this edited collection focuses on everyday life. An introductory example will serve to explain the importance of this domain. The rapid urbanization process faced by most cities in Latin America has created cities that grow faster than their ability to provide adequate infrastructure. Failure to provide suitable mobility solutions for a diverse and growing urban population becomes an increasingly pressing issue as people switch from predominantly pedestrian and public transport mobility patterns to private transport, and from informal responses to collective transport. This trend generates great levels of congestion, environmental concerns in terms of air pollution, poor health, contamination, and high energy consumption, as well as potentially widening social inequalities, all of which question the future sustainability of such trends.

To deal with this, the main trends in formal public transport policy include the implementation of BRT (Bus Rapid Transit) systems and metro extensions. For private transport, urban highways are still seen as a panacea. Such developments are increasingly accompanied by Intelligent Transportation Systems that allow for the dynamic control and operation of transit systems and include automatic vehicle locators, centralized vehicle control, integrated signal control, automatic fare collection and real-time passenger information systems. However, almost irrespective of the financial and human resources that are invested in such schemes, the supposed benefits of these innovations rarely translate seamlessly or as intended into better transport for the majority of the population.

Increased use of technology in transport interventions is often implemented with an apparent neglect for citizens' desires or expressed needs, or their willingness and capacity to cope with change. Transport planners and technicians invariably see technology as offering a universal solution to all mobility problems. However, technological innovations are often inadequately aligned with the urban population and their mobility processes, which include the social, cultural, economic, political, physical, and networked implications of moving in and through metropolitan areas. This is often because technicians' conceptions of mobility are, to a greater or lesser extent, still reliant on models which privilege origin and destination attributes above the many other processes and implications of such movement.

Technology inevitably shapes our world and impacts upon our daily decisions, but the extent to which technology shapes our lives greatly depends on our individual ability to access it. What happens when technological advances surpass our ways of coping with them? What happens when technological innovations, which are considered suitable for certain cities, are implemented without an adequate adaptation to local needs? This has serious implications at various levels. There are cities that have minimal water, sewage, and electricity coverage; and deficient road pavement, yet international firms lobby to implement costly underground metro systems whose costs may exceed the yearly budget of local administrations. There are cities that attempt to implement automatic fare payment systems, without adequately installing card-charging systems. There are cities that introduce intricate modelling systems, without understanding that it is human beings that pass through the systems' nodes, and thus the predictive model might not reflect reality. There are cities whose new transport systems do not consider local idiosyncrasies, the implications of which only begin to manifest themselves after the system has been constructed, often controversially, and invariably late and/or over budget.

This is not to say that technological advances are not suitable for use in cities in developing countries, or anywhere else for that matter—far from it. However, cities must consider their local needs and their specific cultural, social, economic, and physical environments in order to introduce technological innovations that might have important consequences for everyday life and urban interaction. This everyday life approach to technological innovations that affect mobility practices is central to the contributors to this book, and is especially prominent in the chapters included in part three. Take, for instance, how telecommunication companies appear to be out of touch with the everyday needs of many mobile phone users in Canada—as discussed by Shepherd and Shade. Consider the utility of such an everyday perspective for better business, better planning, and better regulation, in this and other contexts.

The "everyday" can be understood as events which are otherwise imperceptible, irrevocably lost, and which are perhaps the most truly personal. These hidden aspects refer to those "secret" parts of people's lives that are often ignored or misjudged by urban research and practice, as urban and transport studies often see everyday life as unproblematic or simply invisible in their analyses. Uncovering those aspects, which can remain hidden by abstract quantitative analysis or by qualitative perspectives that enquire on broader understandings of reality, is urgent in mobility studies today. A good example of the potential of such research for newer, more nuanced theorization comes from Conley's chapter, in which automobility is understood less as a macrosocial phenomenon, and more as an intimate world enacted by mundane, but yet very meaningful, interaction rituals.

In a reflexive manner, everyday life is constantly changing the lives of urban dwellers in the same way that they change everyday life. The experience of daily living—the quotidian, the daily routines which may appear as insignificant— are in fact at the core of what we do, who we are, how we express ourselves. Thus, as space around us changes, so does our everyday experience of it. Some changes occur slowly and we seldom notice them until we make them part of our daily practices: they become an accumulation of small changes that, when we suddenly become aware of them and attempt to look back or return to them, the change becomes insurmountable. Others are quick and have instant impact and force us to adapt our daily living accordingly or resist them. It is in the everyday that changes are recognized and perceived, and that the potential for change can be found, thus the need to recognize the importance of grasping its changing logic.

Lefebvre argued that to "reach reality we must tear away the veil, that veil which is forever being born and reborn of everyday life, and which masks everyday life along with its deepest or loftiest implication" (Lefebvre, 1991, p. 57). He insisted on the need to see the activities that might seem insignificant, for instance. It is in the multifaceted, multitasking moments when everyday life becomes the most vivid or tangible, when most people find themselves living more than one life. It is precisely when a person is trying to be, for instance, simultaneously a mother, a wife, and a worker, that the experience of everyday life provides an important view to the complications or ease people experience throughout the day.

The conception of everyday life practices as separate sectors and dichotomies needs to be overcome, in order to see individuals' experiences within a number of spheres and across spatial scales. Hence, adapting tools to capture these experiences becomes a challenge, as this is done under the full knowledge that the everyday is "always going to exceed the ability to register it" (Highmore, 2001, p. 3). The everyday is, then, both a perspective and a question of methodology; of how to study living. Take, for instance, Salazar's chapter, in which we learn about the dispositions of Chileans towards global mobility, and, simultaneously, gain insight into the image of Chile in global travel discourses. It is precisely at the everyday level of almost unspoken attitudes that historical discourses, geopolitical dynamics, mundane interactions, and deeply-seated values intersect—challenging our very epistemologies.

Another issue of everyday life relevant for the analysis of technological innovations refers to comprehending the everyday as a place where conflict can be found, since everyday life can both hide and make vivid images of social differences (Highmore, 2001) in positive or conflictive ways. As Jirón shows in her chapter this requires questioning the transparency of everyday life and exposing it as a problematic and contested terrain, where ready-made meanings are not easily traced and social power relations take place in terms of spatial and economic struggles, negotiations, transformations, resistances, and differentiated experiences. Through this lens we see how difference is crucial in everyday urban analysis, how one person's experience could be so different from another's, regardless of events unfolding under seemingly similar contexts. When seen from a single perspective, analysis often hides unknown aspects of everyday life which could be recognized as being an essential part of the way places are produced, reproduced, and especially, lived.

For instance, the experience of everyday life has often been interpreted in gendered ways: the feminine has been linked to the daily rituals of private life carried out within the domestic sphere traditionally presided over by women; the masculine exists in the public spaces and spheres dominated especially, but not exclusively in modern Western societies, by men. This gendered division is not very useful, as both men and women use both private and public spaces, including public transport, in differentiated manners. A richer analysis would involve looking at how men and women experience the everyday life differently in such areas, and how gender relations, as well as other social relations, affect everyday experience and generate differentiated experiences. Although everyday life is marked by difference, diversity in the experience of everyday life (most obviously noted by class, gender, age, race, sexuality, etc.) can be seen as positive for place-making; however, this experience can also be negative and be the cause of separation, division and conflict. Analysis need not stop at the recognition of difference; it also needs to explain it and analyze its consequences and implications.

In line with this, de Certeau (1984) criticized the forms of power exerted by rational reasoning, including urban and transport planning, which seek to construct a totally controlled space, a site where everything can be rationally calculated and ordered. In practice, the exercise of technocratic reason excludes everyday practices and discourses that fail to conform to this model of abstract rationality, "thereby expunging difference or otherness" (Gardiner, 2000, p. 167). As examined in part three, but more in general throughout this volume, understanding how daily encounters, negotiations, or transformations take place is part of what unveiling everyday practices of mobility is about, and this is central when technological innovations are introduced to everyday living.

References

Adey, P. (2006). If mobility is everything then it is nothing: Towards a relational politics of im(mobilities). *Mobilities, 1*(1), 75-94.
Adey, P. (2009). *Mobility*. London, UK: Routledge.
Canzler W., Kaufmann, V., & Kesselring, S. (Eds.). (2008). *Tracing mobilities: Towards a cosmopolitan perspective*. Aldershot, UK: Ashgate.
Cresswell, T. (2001). The production of mobilities. *New Formations, 43,* 11-25.
Cresswell, T. (2006). *On the move: Mobility in the modern Western world*. New York, NY: Routledge.
de Certeau, M. (1984). *The practice of everyday life*. Berkeley, CA: University of California Press.
DeLanda, M. (2006). *A new philosophy of society: Assemblage theory and social complexity*. New York, NY: Continuum.
Dovey, K. (2010). *Becoming places: Urbanism/architecture/identity/power*. London, UK: Routledge.
Easterling, K. (2011). Fresh field. In N. Bhatia, M. Przybylski, L. Sheppard, and M. White (Eds.), *Coupling: Strategies for infrastructural opportunism* (pp. 10-13). New York, NY: Princeton Architectural Press.
Farias, I., & Bender, T. (Eds.). (2010). *Urban assemblages: How actor-network theory changes urban studies*. London, UK: Routledge.
Gardiner, M. (2000). *Critiques of everyday life: An introduction*. New York, NY: Routledge.
Gehl, J. (2010). *Cities for people*. Washington, DC: Island Press.
Hall, E. T. (1990). *The hidden dimension*. New York, NY: Anchor Books.
Hannam, K., Sheller, M., & Urry, J. (2006). Editorial: Mobilities, immobilities, and moorings. *Mobilities, 1*(1), 1-22.
Highmore, B. (2001). *Everyday life and cultural theory: An introduction*. New York, NY: Routledge.
Hodge, G. (2008). *The geography of aging: Preparing communities for the surge in seniors*. Montreal, Canada: McGill-Queen's University Press.
Ihde, D. (1990). *Technology and the lifeworld: From garden to earth*. Bloomington, IN: Indiana University Press.
Ihde, D. (2002). *Bodies in technology*. Minneapolis, MN: University of Minnesota Press.
Jensen, O. B. (2007). City of layers: Bangkok's Sky Train and how it works in socially segregating mobility patterns. *Swiss Journal of Sociology, 33*(3), 387-405.
Jensen, O. B. (2009). Foreword: Mobilities as culture. In P. Vannini (Ed.), *The cultures of alternative mobilities: Routes less travelled* (pp. xv-xix). Farnham, UK: Ashgate.
Kaufmann, V. (2002). *Re-thinking mobility: Contemporary sociology*. Aldershot, UK: Ashgate.
Kellerman, A. (2006). *Personal mobilities*. New York, NY: Routledge.
Larsen, J., Urry, J., & Axhausen, K. (2006). *Mobilities, networks, geographies*. Aldershot, UK: Ashgate.
Latour, B. (2005). *Reassembling the social: An introduction to actor-network-theory*. Oxford, UK: Oxford University Press.
Lefebvre, H. (1991). *The production of space* (D. Nicholson-Smith, Trans.). Malden, MA: Blackwell.
Merleau-Ponty, M. (1962). *Phenomenology of perception* (C. Smith, Trans.). New York, NY: Routledge.

Mitchell, W. J. (2004). *Me++: The cyborg self and the networked city.* Cambridge, MA: MIT Press.
Mumford, L. (1934). *Technics and civilization.* San Diego, CA: Harvest Books.
Sheller, M. & Urry, J. (2006). The new mobilities paradigm. *Environment and Planning A, 38*(2), 207–226.
Thrift, N. (2008). *Non-representational theory: Space, politics, affect.* London, UK: Routledge.
Urry, J. (2000). *Sociology beyond societies: Mobilities for the twenty-first century.* London, UK: Routledge.
Urry, J. (2007). *Mobilities.* Malden, MA: Polity.

Part I:
Mobility, Technologies, and the Assemblage of Place

2

Virtual Caribbeans: Edens, Economies, Elsewheres

Mimi Sheller

> With the dismantling of [the BBC's] Caribbean Service...what little that was left that represented and conveyed a regional broadcast perspective will disappear forever. As a consequence the sole vehicle offering the region the chance to hear on a daily basis about events from a broader perspective and sometimes hold politicians to account, will be no more and leading figures in public life will find it virtually impossible to present their views to a region wide radio audience.... In an unspoken way the Caribbean service indirectly supported the sense of region, albeit from afar. That it is going now just as the Caribbean's commitment to regionalism is fading suggests either a lack of political awareness or interest in London. (Jessop, 2011, p. 1)

With the March 2011 demise of the BBC World Service in the face of severe UK budget cuts, political commentator David Jessop suggested that so grave an injury has been done to the existence of the Caribbean that it may simply fade away or dissolve as a region. By the end of March 2011 the Caribbean will "virtually" cease to exist, having only been a virtual entity in the first place, apparently summoned into being by BBC radio broadcasts. As an archipelago, the Caribbean region is an assemblage of parts that are not only physical, but also imagined, symbolic, social, and communicative. Archipelagos

> are inventions whose validity and usefulness is contingent on the dynamics of their formations and the particulars of their contexts. Archipelagos, in other words, are not essential properties of space but instead are fluid cultural processes dependent on changing conditions of articulation or connection. (Vannini, Baldacchino, Guay, Royle, & Steinberg, 2009, p. 124)

If Caribbean metageography is grounded in particular ways of thinking and of sensing the material and social world (ranging from New World discovery narratives to regional radio broadcasts), and stored as bodies of knowledge and praxis (such as botanical collections, plantation manuals, or travel guides), as Vannini et al. suggested, then the archipelago functions as a "particular constellation of articulations that selects, draws together, stakes out and envelops a territory that exhibits some tenacity and effectivity" (Vannini et al., 2009, p. 124). While one might take issue with Jessop's presumption of the power of British radio to hail a region into being (Althusser's "hey, you!"), it is nonetheless important to ask in what ways a Caribbean "sense of region" is either fading, or perhaps being reconfigured into different kinds of assemblage, and what role new communication technologies might play in such processes.

Assembling the Virtual Caribbean

In *Consuming the Caribbean* (Sheller, 2003), I followed the lead of Sarah Franklin, Celia Lury, and Jackie Stacey in their study, *Global Nature, Global Culture*, to think of the Caribbean as an effect, a fantasy, a set of practices, and a context (terms which they use to describe "global nature" as culture/cultured). This is not to suggest that the Caribbean is "illusory, immaterial, or a matter of ideas and imagination alone—far from it" (Franklin, Lury, & Stacey, 2000, p. 5), but rather to begin to grasp how nature is both "seconded" and made "second nature." That is to say, it is both enlisted to do certain kinds of work of imagination (denaturalized), and then made to seem original and untouched, a kind of encultured nature (renaturalized). The virtual and the real (or realized) mirror each other and enable each other, becoming entangled, with each being a precedent for the other. Insofar as the Caribbean is both denaturalized and renaturalized as a fantastical "natural paradise," it is a perfect example of the shimmering simultaneity that Homay King has identified in which virtuality "invokes existence and non-existence, reality and unreality" (King, 2011, n.p.). It is because of this ontological oscillation that we can ask how Caribbean virtuality was enacted in the past, and might be changing under current conditions.

The concept of "virtual islands" is one way to think about Caribbean spatiality and its new metageographies, but first I want to distinguish between virtual worlds that exist only on the Internet, which are *not* the subject of my inquiry, versus those that straddle what I call "hybrid spaces," drawing on recent work on mobile gaming and mobile social networks (de Souza e Silva, 2009; de Souza e Silva & Sutko, 2008, 2009). In the largely digital realm of "cyberspace" there are:

1. The virtual worlds formed by members of Caribbean diasporas who create home pages, electronic mailing lists, and online communities of various kinds that re-create and extend Caribbean identities into cyberspace.[1]

2. Caribbean governing agencies, business interests, and civil society organizations that create parallel online representations of their interests, institutions, and organizations, hence a kind of virtual extension of the physical space of the island across the "infosphere" (McDowell, Steinberg, & Tomasello, 2008).

3. Media representations of the Caribbean in fictional forms, such as the circulation of cinematic images like the Pirates of the Caribbean movie series (actually filmed in the Windward Island of Dominica), computer games like Tropico in which players build simulated Caribbean island societies (Edward, 2006), or the use of simulated islands more widely in multiplayer virtual worlds like Second Life, creating a universe of virtual islands in informf ormation territories.

4. The vast realm of tourist marketing imagery, which circulates representations of imagined and imaginary islands (beach, palm trees, sun, and sex) across a range of print, broadcast, and electronic media.

I do not directly include any of these here—except insofar as they directly impact on and reshape the physical world. My focus, instead, falls on the interaction of print and digital representations of space with physical practices of space to form "Virtual Caribbeans."[2] Dodge and Kitchin (2001) suggested that "an *essential* element in understanding ICTs [information and communication technologies] and cyberspace is a comprehension of how they are transforming, and creating new spatialities, spatial forms and space-time relations" (p. x). Building on Michel Foucault's archaeology of knowledge, Bruno Latour's actor-network theory, and Nigel Thrift's non-representational theory, a "neo-materialist" philosophy of the social sciences engages with the interface between social and physical systems and human and non-human assemblages.

Within the emerging field of software studies it is also recognized that the "materiality" of software is operative at many different scales, including as "an embedded part of socio-technical infrastructures" (Fuller, 2008, p. 5). And such hybrid spaces are crucial to understandings of locative media in relation to territory:

> All "territory" owes something to the processes of the virtual.... All territory is made of information; it is not mere data but formed, ordered and willed animal territory.... The informational territory is not cyberspace, but territory in a place formed by the relationship between the physical dimensions of territorialities and the new electronic flows, creating a new form of territorialization. (Fiorelli, Lemos, & Shields, 2009, n.p.)

Caribbean territories thus owe something to the processes of the virtual, and we might even suggest that as an archipelago of islands assembled into a temporary regionalism they are especially implicated in virtual processes. Moreover, "virtuality—as the capacity of communicative technologies to constitute rather than mediate realities and to constitute relatively bounded spheres of interaction—is neither new nor specific to the Internet" (Fuller, 2008, p. 6).[3] As Daniel Miller and Don Slater (2000) crucially argued in their foundational ethnographic study of the Internet, "we need to treat Internet media as continuous with and embedded in other social spaces" (which, in their case, happened to be in Trinidad). Drawing on the science studies of technology of Bruno Latour as well as traditions within the anthropology of material culture, they argued that Internet media "happen within mundane social structures and relations that they may transform but that they cannot escape into a self-enclosed cyberian apartness" (Miller & Slater, 2000, p. 5). Cyberspace is not a place apart from the rest of social life; but beyond that, one could add, the rest of social life also engages virtual processes that are not part of cyberspace.

Better understandings of the complex varieties of Virtual Caribbeans are crucial to understanding both the historical formation of the modern Caribbean archipelago and the contemporary forms of deterritorialization and reterritorialization brought about by cyberspatial restructurings of states, economies, cities, citizenries, and societies. Paying attention to virtual sovereignty and place-making—suspended between the real and the imaginary—offers a key terrain (or a virtual foothold) for thinking about the contemporary rescaling of Caribbean island territories. This includes their supranational "extended statehood systems" (de Jong & Kruijt, 2006) and their subnational "offshoring strategies" (Baldacchino, 2010), which create subnational island jurisdictions that arrange special relationships with larger metropolitan powers, pioneering new forms of sovereignty and government. This work in progress investigates the scattered geography of virtually augmented Caribbean island space in order to show how software and information and communication technologies are assembling islands in the Caribbean into new hybrid configurations of mobility, territoriality, and governance.

I build on previous work (Sheller, 2004a, 2004b, 2007b, 2009a, 2009b) to show more exactly how the use of software, the invention of cyberspatial property, and the creation of various kinds of virtual islands are recoding (in terms of both legal codes and software codes) Caribbean space, with crucial implications for regional identities and national citizenries. The ultimate aim is to develop a critical analysis of the future of Caribbean civil societies, states, and survival in an age of virtual reconfigurations of territorial spatiality and of island economies and geographies. In the following three sections I explore three aspects of the mixed virtual realities of Caribbean islands, which I refer to as virtual Edens, virtual economies, and virtual elsewheres. In each case not only is the line between the real and the virtual blurred, but the action takes place across that line, on that line, or between the lines—with the indistinction being the fulcrum on which futures balance.

Virtual Edens

There is no "primal nature" in the Caribbean, not only because so much of it has been transformed by human interventions (species introductions and extinctions, deforestation, population impact, coral bleaching, etc.), but also because every aspect of our perception of the Caribbean assemblage is laden with historical allusions, second order simulations, and hackneyed repetition. There is no need to write the story of the Caribbean as Garden of Eden (with or without the serpent), as Paradise (with or without threatening cannibals), or Playground (with or without pirates), because we already know it. With every palm tree, every rum drink or tin-pan melody we recognize it, sober or drunk, as a party theme, a cheap holiday, a casino décor, or a vacation package. The Caribbean is always in drag, touting its own image. Nevertheless, I want to pause a moment to consider how textual representations of island paradise interacted historically with the production of island spatiality, for this is the first (predigital) phase of Caribbean informational territorialization.

With regard to the 17th to 19th centuries, Richard Grove argued that "the commercial and utilitarian purposes of European expansion produced a situation in which the tropical environment was increasingly utilized as the symbolic location for the idealized landscapes and aspirations of the Western imagination" (Grove, 1995, p. 3). On the other hand, one could flip that around to argue that the idealization of the tropical island landscape in the Western imagination produced a situation in which mythic representations gave way to more factually grounded geographical texts, and these texts, maps, charts, and seafaring guides instigated and enabled Europeans to expand their commercial and utilitarian projects. Recent studies of geographical writing focus on the effective role of texts, maps, sea charts, and books in shaping "epistemes," forming economies, and consolidating spatialities (Brotton, 2004; Ogborn, 2007; Saldanha, 2011). Thus, Saldanha argued, it was the "sudden eruption of a new but diffuse understanding of imminent possibility" (2011, p. 172) embodied in Linschoten's *Itinerario* (1596), that impelled the Golden Age of Dutch navigation. His argument is suggestive of the role of geographical texts in imagining virtual (imminent) worlds that invite exploration, and (much like the BBC Caribbean Service) hail distant lands as regional entities that can become "real" only through an act of physical and imaginative connection (an itinerary).

The Edenic island is a Western trope dating back at least to the myth of the lost island of Atlantis, recounted by Plato, taking off with Renaissance-era *isolarios*—atlases of both real and imagined islands—and burgeoning into a whole series of classic Western texts: Christopher Columbus's journals, Sir Thomas More's *Utopia*, William Shakespeare's *The Tempest*, Daniel Defoe's *Robinson Crusoe*, Robert Louis Stephenson's *Treasure Island*, and other such island imaginaries (see Sheller, 2003). As Grove (1995) and others have convincingly shown, Caribbean discovery narratives already drew on a range of precedents such as the biblical Garden of Eden, the classical garden of the Hesperides, and the Renaissance botanical garden, which was itself derived from Middle Eastern models. Columbus may even have believed that the Garden of Eden actually existed in the extreme Orient, but like the ancient Greek Elysium was reached by sailing West and South (see Gómez, 2008); that would explain why he took Luis de Torres, a scholar of Hebrew, Arabic, and Chaldaic with him (Prest, 1981, p. 31), and believed until the end of his life that he had discovered islands near "Cipangu" (Japan). The West Indies, of course, are named after the East Indies, thus they already represent a kind of displaced geography of the imagination and a misrecognition that "orients" the Caribbean in particular ways.

The field of island studies has drawn attention to islands as "sites of innovative conceptualizations, whether of nature or human enterprise, whether virtual or real" (Baldacchino, 2006, p. 3). Over the centuries travellers have continued to see the Caribbean through the eyes of others, in a kind of "travelscripting" guided by previous texts, which "produces a serialized space of constructed visibility that allows and sometimes even requires specific objects to be seen in specific ways by a specific audience.... [T]hey all carry within them traces of the physical movement of embodied subjects through material landscapes" (Gregory, 1999, pp. 116–117). Traversing the Caribbean landscape, the traveller constructs it virtually, composing plants, people, buildings, and light into preconceived scenes/scenery based on books already read, images already conjured.

As one reads travel writing spread across centuries, the Caribbean invariably and repeatedly "seemed to realize the youthful visions of imagination."[4] A Scottish traveller to Barbados in 1802, McKinnen saw before him a picture that was a realization of childhood enchantment, the culmination of the very emotional investments that had already brought him to travel to this far-away place. The land itself is scripted into performing as fantasy. A traveller in Grenada in 1833 observed that from "the parterre before this charming dwelling a beautiful map was spread out before us"[5] (Alexander, 1833, Vol. I, p. 244). The land becomes the map, the map a Borgesian virtual land. English novelist, Charles Kingsley, in the 1870s imagined the first discoverers of the West Indies believing themselves "to have burst into Fairy-land—to be at the gates of the Earthly Paradise." Another writer in 1900 related that "Here, for the first time, was the tropical beach! How often, from childhood, I had tried to picture it from Kingsley's vivid descriptions or the histories of the early explorers." One time period is telescoped into another, as we read Columbus via McKinnen via Kingsley via Paton via Fermor via Walcott.[6]

For such travellers the entire archive of travel writing informed their experience of the Caribbean; "real" experiences were always just a reflection of the vivid Caribbean in their imaginations. History blurs together timelessly through the centuries, as promoted, for example, in a series of advertisements for Alcoa Steamship Company cruises in the 1940s. One orientalized the Caribbean as a place where "East meets West," with the image of an Indo-Trinidadian (see Figure 1); another exoticized it as a "parade of ports" that is "rich in turbulent history," using an image of a "Carib Indian" mother and child (see Figure 2).

Time seems not to matter in these virtual worlds, except as repetition, or stops on an itinerary, and landscapes, flora, fauna, and people are assembled as a curated experience.

Figure 1. Alcoa Sails the Caribbean Advertisements, Holiday Magazine, August 1941

Figure 2. Alcoa Sails the Caribbean Advertisements, Holiday Magazine, February 1942

Today the means of communication, representation, transportation, and circulation have all changed, but the template of repetition continues to assemble the Caribbean into an archipelago of imminent experience and potential profit. Informational space, political space, and tourist space in the Caribbean are converging in new fantasies of mobility, accessibility, and island paradise that are updating these long-standing Western fantasies of the imagined island as an alluring and fascinating *tabula rasa* (Gillis, 2004). Caribbean islands have become sites of cyberspatial innovation and fantasy. Software is recoding and rescaling island space, assembling Caribbean islands into new configurations of territoriality and governance while generating new kinds of atmospheres, new sensuously engaged imaginaries of place, landscape, nature, and bodies. Travel and leisure destinations, especially in the Caribbean, are being disembedded from national territories and repackaged as local enclaves that are hyperconnected to global metropolitan transport, media, and data flows, suggesting a kind of materialized virtuality or, more radically, a virtualization of the material realm.

Virtual Economies

Virtual economies are not new to the Caribbean, and may even be thought of as foundational. In his work on how the slave trade established a "new discourse and configuration of credit" in the early 18th century, financial historian Carl Wennerlind examined how promoters of the South Sea Company, made use of "propaganda writers [who] worked tirelessly to ensure that the public visualized the trans-Atlantic slave economy as an inexhaustible fountain of riches" (2011, p. 198). While many have focused on the financially catastrophic bursting of the South Sea bubble in 1719–1720, Wennerlind suggested that the earlier 1710 innovation in credit was crucial. England's massive national debt was miraculously converted into Company shares by piquing the public imagination with interest in the extraordinary profits to be made in the slave trade. Fictional accounts (such as those by Daniel Defoe), pamphlets, newspapers, ballads, and other popular forms were central to this confidence scheme, built on the "formation of the public's social imaginary of the Atlantic slave trade" (p. 197). "The term social imaginary is here used to refer to a collective vision, perception, or understanding of a world unobserved by most English people" (Wennerlind, 2011, p. 315; cf. Sherman, 1996). In so far as credit and finance were dependent on imagination, credit, according to Catherine Ingrassia (1998), "was not a 'real' event; rather it was a phenomenon that could be known primarily through print sources" and "recreated imaginatively in the mind of the investor" (p. 6). The decision to invest in a stock or a bond (or a slave-trading voyage) necessitated the participation in "an imaginatively based narrative" (p. 7), which drove the even larger project of creating a new economy. The social imaginary of credit, then, is another kind of virtual (yet effective) world, a financial fiction that drove the bitter materiality of the slave trade and hence the making of the modern Caribbean.

So the slave trade, and with it the economic basis for the Atlantic plantation economies, hinged on an act of imagination, a wager, a collective belief in the real profits to be made from a still-in-formation virtual world. Wennerlind drew on Marx to describe this as a kind of fetishism, but one that differs from commodity fetishism:

> The abstraction, or fetishism, facilitated by credit differs in important ways from that which Marx ascribed to money. If money allows people to disregard the origins of value, credit on the other hand necessitates a careful construction of a social imaginary of the reality within which future values will be produced. Money frees people from thinking about the past conditions of production, while credit necessitates that the future conditions are carefully considered and vividly imagined. In other words, money and credit enable different practices of abstraction and different kinds of fetishism. (2011, p. 230)

And, I would argue, different temporalities of virtuality. The futures markets on which credit hinges were dependent on remaking Caribbean islands as plantation societies through the importation of capital, tools, food, and, above all, enslaved laborers. Rather than abstracting labor into the commodity fetish, to be exchanged for money, credit turned real human beings into commodities, to be exchanged for money and turned into profit for others. The material world and social relations were rearranged by the investments that drove the slave ships, while the system of slavery transformed the people of Africa and the islands of the Caribbean into something real yet unreal, human yet not-human, modern yet not-modern; they occupied a "discrepant modernity," as James Clifford called it, or a "Savage Slot" that was in, but not of, the West (Trouillot, 2003).

Today offshore economies in the Caribbean are areas in which sovereignty and reality are also abstract, imaginary, and, in a sense, virtual: "an 'in-between' juridical realm where states are able or willing to apply only a certain degree of regulation" (Palan, 1998, p. 637). Offshore economies—including tax havens, free-trade zones, export-processing zones, flags of convenience, Internet business, shell banks, and so on—are not evidence of globalization undermining state power, but are instead part of a larger process of the radical redrawing of state sovereignty (Cameron & Palan, 2004). More broadly speaking, "offshoring processes" include the engineering of extra-national, quasi-autonomous territories: detention camps like Guantánamo Bay, military bases, world heritage sites, and otherwise autonomous jurisdictions that are lifted out of normal sovereignty (Baldacchino, 2010). Companies, banks, and investments exist in a kind of virtual reality, legally located in the Caribbean, but detached from any physical region; at the same time they enable property investors to buy up extensive coastlines of the Caribbean archipelago to be dedicated to tourism and non-resident property markets.

The movement of money is, ironically, one of the few sectors in which some small non-independent Caribbean territories have enjoyed a comparative advantage, in part due to their amorphous political status as insular enclaves that are in between dependency and full sovereignty. Several U.K. dependencies or former colonies are well-known tax havens whose unregulated banking greases the wheels of global financial velocity: Anguilla, Antigua, Belize, Bermuda, the British Virgin Islands, the Turks and Caicos, and Grenada. The deregulation of these spaces has facilitated new "state fictions," including the flourishing of new forms of cyberspatial property (such as private islands bought on timeshares and accessed by nonresident owners via online, charter, private-jet services like NetJets) enabled by Internet transactions in which property is undergoing virtualization thanks to more fluid financial transactions and new software-supported infrastructures (see Maurer, 1995, 2002).

In the 21st century, Caribbean space is being socially and politically produced under new conditions of commercialized sovereignty, virtual cyberproperty, and fictional residency. Software-supported air mobility, freight logistics, data processing, property development, building design, marketing, banking, travel, and surveillance together enable a disembedding of island space from structures of local governance and territoriality, and a disassembly of the archipelago as a networked group of islands. As I have examined in a case study of Dellis Cay in the Turks and Caicos (Sheller, 2009a, 2009b), new forms of infrastructural exclusivity, computer-aided design, media-savvy, web-based property marketing, and uneven forms of software-sorted mobility underwrite proprietary regimes that assist in channelling who has access (or does not) to various kinds of real estate and residency rights. Offshore luxury homes in tax havens create fictions of residency and domicile that allow individuals and corporations to hold juridical residency without actually being present; ICT connectivity creates the conditions for such punctuated presence. Software-supported architecture and "infrascapes" (infrastructure landscapes) assemble and put into practice these new hybrid spatialities, permitting new kinds of tourism performances, resort development, and property ownership that create new hybrids of the virtually present island.

Such deregulated or weakly regulated territories "are all part of a worldwide network of essentially marginal places which have come to assume a crucial position in the global circuits of fungible, fast-moving, furtive money and fictitious capital" (Roberts, 1994, p. 92, as cited in Warf, 1995/2005, p. 359). At least, they did, until a run of bad fortune clipped their wings. First, crackdowns on the narcoeconomy (War on Drugs) and on terrorist financial networks (War on Terror) brought the attention of the OECD regulators. Then there was the collapse of Texas billionaire, Robert Allen Stanford's International Bank Ltd., based in Antigua, in a massive Ponzi scheme, not unlike the Bernard Madoff fiasco. This was followed shortly after by the British government's takeover of the ignominious Turks and Caicos government following charges of corruption and incompetence linked to property speculation. As in the days of piracy and freebooting, the Caribbean has come to be associated with interruptions of the normal flows of capital and sovereign power, as well as with forms of corruption, smuggling, drug running, gun trading, and money laundering, which allegedly subvert (yet in so doing support) the formal regulated economy. There is a blurring between legality/illegality and policing/crime as criminals, gangs, and fraudsters seem to infiltrate the government in many states, and state power hesitates to bridle known outlaws. State functionaries are themselves above and beyond the law; "real" economies are clouded by underground "grey" economies.

Some theorists connect the offshore with Giorgio Agamben's notion of the state of exception, in which xenomoney and xenospace are traced back to European travel and mobility, territorial fictions, and money as fiction. Angus Cameron drew on Georges Bataille and Jean-Joseph Goux, for example, in a project on the offshore Bahamas, to describe the offshore as excessive, erotic, exclusionary.[7] The offshore becomes the virtual location in which the "accursed share" is consumed in a conflagration of wasted profit, as Bataille (1967, 1989) described:

> if the system can no longer grow, or if the excess cannot be completely absorbed in its growth, it must necessarily be lost without profit; it must be spent, willingly or not, gloriously or catastrophically.... Minds accustomed to seeing the development of productive forces as the ideal end of activity refuse to recognize that energy which constitutes wealth must ultimately be spent lavishly (without return), and that a series of profitable operations has absolutely no other effect than the squandering of profits. (1989, pp. 21–22)

Here the potentialities of virtual capital are perversely consumed in destructive processes rather than productive ones, in which human labor, ecologies, and lives are laid waste by an excessive expenditure of capital. And it is in such virtual economies that entire populations can be left to fend for themselves, falling into a kind of limbo in which no state claims responsibility for their existence or well-being. When 300,000 people were obliterated in 32 seconds and 1.5 million left homeless in Port-au-Prince during the earthquake of January 12th, 2010, the rest of the world was momentarily awakened to a reality that had been an abstraction, two hours from Miami, but a million miles away in terms of the accursed metageography of the Savage Slot.

Virtual Elsewheres

In April 1999, two US F-18 airplanes dropped two five-hundred pound bombs outside their target area, the weapons testing range on Vieques Island, Puerto Rico, killing David Sanes, a civilian security guard, and injuring four others. This provoked an existing resistance movement to establish civil disobedience camps inside the bombing range, and in February 2000 the largest mass demonstration in Puerto Rican history was organized in San Juan, the Marcha para la Paz de Vieques (Baver, 2006; McCaffrey, 2002). Through "new and old media technologies, from internet sites to film festivals, newspapers to television, the civil disobedience campaigns of the late 1990s and early 2000s made the Vieques Libre struggle for peace internationally known" (Obrist, 2005, p. 205), as represented in the "Land Mark" project by Puerto Rico-based experimental performance artists, Jennifer Allora and Guillermo Calzadilla (who addressed environmental and social justice issues surrounding the U.S. Navy's appropriation of Vieques Island as a weapons-testing range). This hastened the process of military withdrawal from the island (from May 2001 to May 2003). After 60 years of U.S. Navy presence in Vieques—which allegedly included testing live ammunition, depleted uranium shells, napalm, and germ and chemical warfare—the resistance movement had driven out the Navy (see Baver, 2006).

Allora and Calzadilla had participants leave footprints on the beaches of Vieques, which they then recorded, as part of a project that reflected upon the relation between land and the marks upon it, the landscape and the power to landmark or mark land. At the same time, however, prevailing military practice had already turned towards discourses of "precision bombing" supported by global positioning satellite (GPS) technologies (Kaplan, 2006). Well before the current public debates over the use of remote-controlled predator drones to kill Taliban enemies in the remote hills of Pakistan, Kaplan argued that the rise of GPS and global information systems (GIS) have supported changing cultural practices of targeting within the military-industrial-media-entertainment complex. By the 1990s "phantom" bombing targets were being simulated as imaginary islands and U.S. Navy weapons training in the Caribbean had begun to use Virtual At-Sea Training (VAST),[8] a program that was accelerated in 2002 after the decision to pull out of Vieques. The real island was no longer needed, having been replaced by a virtual target, so the resistance movement met with little resistance.

Most of the U.S. Navy land was returned not to the government of Puerto Rico or to the municipality of Vieques, however, but to the U.S. Federal government. Large portions were designated as a nature reserve and wildlife refuge under the direction of the U.S. Fish and Wildlife Service (Baver, 2006).

Presumably, this gesture of natural "conservation" was supposed to create a beneficent image of the military; yet it also served to cover up the damage to the island under the cover of "nature" and benefits to "wildlife." This has assisted the reinvention of Vieques as an "untouched" tourist paradise (see Sheller, 2007a). Virtual targets replace real bombing ranges, only to be replaced with tourist imaginaries which reinvent the natural paradise of the Caribbean Eden. Once again the circulation of information and representations instigates actions that enable new markets and social relations to emerge, creating new geographies and spatial relations between "here" and "elsewhere."

Such virtual islands are also part of what geographers Stephen Graham and Simon Marvin (2001) described as the "splintering of urbanism" and "unbundling of infrastructure." Virtuality creates filaments of metropolitan connectivity that hop over otherwise abjected and disconnected regions of the world. Caribbean governance is embedded in multi-scalar, transnational systems that unevenly connect some Caribbean regions into metropolitan spatialities even as others are ejected and disconnected. What happens to the real needs, bodies, and rights of those who dwell within that hybrid, informationalized space is a central problem. Virtual realities penetrate or permeate physical spatiality in the fantasy islands of tourist mobility; they assemble space. Yet the disavowed localities of social exclusion and infrastructural disconnectivity are simultaneously cut off and abandoned to violence, informal economies, black markets, and abject forms of survival.[9] The fading sense of a Caribbean archipelago and the emerging deterritorializations and reterritorializations of Caribbean sovereignty and property are thus embedded in social and material realities that are part of longer histories of colonial and postcolonial imaginations of the nation, of the state, and of legal jurisdictions governing the unrealized rights of citizenship, residency, and human status.

But I would be remiss not to recognize the ways in which the unstable fictions of Caribbean virtuality also offer creative leverage for Caribbean artists, writers, and critical theorists. Trouillot (2003) developed the concept of "alter-Native modernities," in which modernity is both altered and alternative in the territories of the non-Western "native." Martinican theorist Edouard Glissant's *Poetics of Relation* (*Poétique de la Relation*) (1997) and *Caribbean Discourse* (*Le Discours antillais*) (1992) also developed a Caribbean theorization of mimesis, "transversality," and "parallactics," all ideas meant to suggest forms of relationality between two places produced through "that intricate doubling that informs postcolonial culture" (De la Campa, 1997, p. 109; see also Glissant, 1992, 1997). Like the doubling of the virtual/real, it is not clear which is the original and which is the copy in these alter-Native modernities. Dwelling in such hybrid spaces can be both productive and profitable.

I want to conclude with the writing of Barbadian poet, Kamau Brathwaite, who subtitled his poem "Guanahani" (the indigenous name for some of the Bahamian islands), "flying over the Bahamas 12 Oct 1492 on AJ 016 over the US Easter [sic] Seaboard of Gauguin."[10] Telescoping past and present, primitive and modern, and superimposing various parts of the planet and their distinctive geographies, the poem is narrated as if looking down at Earth from a spacecraft:

> How come
> along the East Coast of North America
> almost to noon the thin white line of the long beach
> the clouds coming right down on the water like ice-floes
> like thousands of tiny floating islands in an orange tint of water

Brathwaite dealt with elements and geographies in motion and went on to refer to "the ice-floes drifting over the Arctic's infidelity" and the "clouds soon crowding again/strato-cumulus of beginning of the moon from 39,000 feet of the spacecraft." In this dislocated time-space he becomes a "witness" to the "outer wheels/& limits of spiral galaxies triangles parachutes. shapes of magenta/stealth bombers ghosts shrouds/Tibetan journeying spaces of time between magnets & continents/causeways into another continuum. approaching the new life of Eleuthera."[11] Brathwaite projected the Caribbean subject backwards in time and forwards into the Space Age, producing a version of Afro-futurism (Eshun, 1999; McLeod, 2003). We can read this New World, Afro-futurist vision as producing parallel worlds of the possible, autochthonous versions of a Caribbean virtual elsewhere. Through transversal movements across doubled worlds, Caribbean theorists call into question which world is primary and which secondary; and in which temporality the virtual witness is present or absent. They remind us that there are multiple imminent futures in which the Caribbean may or may not virtually survive.

Notes

1. For an excellent discussion of Caribbean bloggers see Paravisini-Gebert and Romero-Cesareo (2011).

2. "Virtual Caribbeans: Conference on Representation, Diaspora and Performance in and on the Caribbean," was held at Tulane University, New Orleans, LA, February 28–March 1, 2008. It suggested that "The definition of the Caribbean as primarily a geographical region is no longer viable. Through the movement of its peoples, cultures, and languages, we

also make or find the "Caribbean" elsewhere" (http://tulane.edu/liberal-arts/spanish-portuguese/events.cfm)

3. Fiorelli, Lemos, and Shields (2009) stated:

> Virtualization means that selected elements of places and selected aspects of place-making and conferring meaning on places and regions are lifted out of the actual world of material sites into the ideal but still real environment of digital media. Place is now much more clearly part material, part virtual and part representational. (n.p.)

4. See in particular Daniel MacKinnen (1804), *A Tour Through the British West Indies, in the Years 1802 and 1803, Giving a Particular Account of the Bahama Islands* (pp. 6, 16).

5. See Captain J. E. Alexander (1833), *Transatlantic Sketches, Comprising Visits to the Most Interesting Scenes in North and South America, and the West Indies, With Notes on Negro Slavery and Canadian Emigration*, Vol. I, p. 244.

6. Charles Kingsley (1873), *At Last: A Christmas in the West Indies*, pp. 26-27; William Agnew Paton (1888), *Down the Islands: A Voyage to the Caribbees*; E. A. Hastings Jay (1900), *A Glimpse of the Tropics, or, Four Months Cruising in the West Indies*, pp. 34-35; Patrick Leigh Fermor, *The Traveller's Tree: A Journey Through the Caribbean Islands*; Derek Walcott, "A Frosty Fragrance," *The New York Review*, June 15, 2000, p. 61.

7. See http://www.goldinsenneby.com/gs/?p=103

8. One of the Navy's most advanced training systems, VAST uses computers that create a virtual island (or any other venue), based on photography of actual locations. Though military personnel involved in the exercise will see the island on their computer screens, their actual target is cordoned off by up to five sonar buoys far out at sea. These buoys pinpoint ordnance hits, calculate where they would have struck the island and provide smoke and explosions to signal the impacts. A predator drone is used to provide a simulated view of the target area. (Adams, 2003, p. 36)

9. Virtual mapping and software support are inherent not only in military power in the Caribbean region, but also humanitarian interventions and post-disaster response and logistics, as seen recently in post-earthquake Haiti (see Sheller, 2010). New forms of aerial vision such as Google Earth are used for disaster response but also offer an empowering gaze, assisted by satellite vision and mobile GPS, which updates the militarized aerial power of the past, described by Caren Kaplan (2006) as a "cosmic view."

10. Guanahani is the name the indigenous people gave to the island in the Lucayan archipelago of what is now the Bahamas, first sighted by Christopher Columbus on October 12th, 1492, and named by him "San Salvador." For more, see Kamau Brathwaite, "Guanahani," in *Born to Slow Horses* (2005, p. 7).

11. Eleuthera is a long thin island in the Bahamas, whose name in Greek means "free"; it was also the site of a U.S. Auxiliary Airforce base and the U.S. Air Force Eastern Test Range (ETR) Tracking Station #4, in the 1960s-1970s.

References

Adams, E. (2003). The Navy's phantom bombing range. *Popular Science, 262*(2), 36-37.
Alexander, J. E. (1833). *Transatlantic sketches, comprising visits to the most interesting scenes in North and South America, and the West Indies, with notes on Negro slavery and Canadian emigration* (2 Vols.). London, UK: R. Bentley.
Baldacchino, G. (2006). Islands, island studies, Island Studies Journal. *Island Studies Journal, 1*(1), 3-18.
Baldacchino, G. (2010). *Island enclaves: Offshoring strategies, creative governance, and subnational island jurisdictions.* Montreal, Canada: McGill-Queen's University Press.
Bataille, G. (1967). *La part maudite.* Paris, France: Les Editions de Minuit, 1967.
Bataille, G. (1989). *The accursed share* (Vol. I; R. Hurley, Trans.). New York, NY: Zone Books.
Baver, S. (2006). 'Peace is more than the end of bombing': The second stage of the Vieques struggle. *Latin American Perspectives, 33*(1): 102-115.
Brathwaite, K. (2005). *Born to slow horses.* Middletown, CT: Wesleyan University Press.
Brotton, J. (2004). *Trading territories: Mapping the early modern world.* Chicago, IL: University of Chicago Press.
Cameron, A., & Palan, R. (2004). *The imagined economies of globalization.* London, UK: Sage.
de Jong, L., & Kruijt, D. (Eds.). (2006). *Extended statehood in the Caribbean: Paradoxes of quasi colonialism, local autonomy, and extended statehood in the USA, French, Dutch and British Caribbean.* Amsterdam, The Netherlands: Rozenberg Publishers.
De la Campa, R. (1997). Resistance and globalization in Caribbean discourse: Antonio Benítez-Rojo and Edouard Glissant. In A. J. Arnold (Ed.), *A history of literature in the Caribbean, Vol. 3: Cross-cultural studies.* Amsterdam, The Netherlands: John Benjamins, (pp. 87-116).
de Souza e Silva, A. (2009). Hybrid reality and location-based gaming: Redefining mobility and game spaces in urban environments. *Simulation & Gaming, 40*(3), 404-424.
de Souza e Silva, A., & Sutko, D. (2008). Playing life and living play: How hybrid reality games reframe space, play, and the ordinary. *Critical Studies in Media Communication, 25*(5), 447-465.
de Souza e Silva, A., & Sutko, D. (Eds.). (2009). *Digital cityscapes: Merging digital and urban playspaces.* New York, NY: Peter Lang.
Dodge, M., & Kitchin, R. (2001). *Mapping cyberspace.* New York, NY: Routledge.
Edward, M. (2006). Simulating the Caribbean: An analysis of the computer game Tropico. In S. Courtman (Ed.), *The Society for Caribbean Studies Annual Conference Papers*, Vol. 7. Retrieved from http://www.caribbeanstudies.org.uk/papers
Eshun, K. (1999). *More brilliant than the sun: Adventures in sonic fiction.* London, UK: Quartet Books.
Fermor, P. L. (1955). *The traveller's tree: A journey through the Caribbean islands.* New York, NY: Harper.

Fiorelli, M., Lemos, A., & Shields, R. (2009). Sur-viv-all: Locative art. *Wi: Journal of Mobile Media*. Retrieved from http://wi.hexagram.ca/?p=47

Franklin, S., Lury, C., & Stacey, J. (2000). *Global nature, global culture*. London, UK: Sage.

Fuller, M. (Ed.). (2008). *Software studies: A lexicon*. Cambridge, MA: MIT Press.

Gillis, J. (2004). *Islands of the mind: How the human imagination created the Atlantic world*. New York, NY: Palgrave Macmillan.

Glissant, E. (1992). *Caribbean discourse: Selected essays* (J. M. Dash, Trans.). Richmond, VA: University Press of Virginia.

Glissant, E. (1997). *Poetics of relation* (B. Wing, Trans.). Ann Arbor, MI: University of Michigan Press.

Gómez, N. W. (2008). *The tropics of empire: Why Columbus sailed south to the Indies*. Cambridge, MA: MIT Press.

Graham, S. and Marvin, S.(2001) *Splintering urbanism: networked infrastructures, technological mobilities and the urban condition*. London and New York: Routledge.

Gregory, D. (1999). Scripting Egypt: Orientalism and the cultures of travel. In J. Duncan & D. Gregory (Eds.), *Writes of passage: Reading travel writing* (pp. 114-150). London, UK: Routledge.

Grove, R. (1995). *Green imperialism: Colonial expansion, tropical island Edens and the origins of environmentalism, 1600-1860*. Cambridge, UK: Cambridge University Press.

Hastings Jay, E. A. (1900). *A glimpse of the tropics, or, four months cruising in the West Indies*. London, UK: S. Low, Marston.

Ingrassia, C. (1998). *Authorship, commerce, and gender in early eighteenth-century England: A culture of paper credit*. Cambridge, UK: Cambridge University Press.

Jessop, D. (2011, January 28). The view from Europe. Retrieved from http://www.caribbean-council.org/?q=ViewFromEurope#The%20View%20from%20Europe

Kaplan, C. (2006). Precision targets: GPS and the militarization of U.S. consumer identity. *American Quarterly, 58*(3), 693-714.

King, H. (2011). *The dream of the virtual: Digital frontiers and the flight from earth*. Unpublished draft presented in the Penn Humanities Forum, February 8, 2011.

Kingsley, C. (1873). *At last: A christmas in the West Indies* (new ed.). London, UK: Macmillan.

MacKinnen, D. (1804). *A tour through the British West Indies, in the years 1802 and 1803, giving a particular account of the Bahama Islands*. London, UK: J. White; R. Taylor.

Maurer, B. (1995). Complex subjects: Offshore finance, complexity theory, and the dispersion of the modern. *Socialist Review, 25*(3-4), 113-145.

Maurer, B. (2002). A fish story: Rethinking globalization on Virgin Gorda, British Virgin Islands. In J. X. Inda & R. Rosaldo (Eds.), *The anthropology of globalization: A reader* (pp. 100-135). Oxford, UK: Blackwell.

McCaffrey, K. (2002). *Military power and popular protest: The U.S. Navy in Vieques, Puerto Rico*. New Brunswick, NJ: Rutgers University Press.

McDowell, S., Steinberg, P., & Tomasello, T. (2008). *Managing the infosphere: Governance, technology, and cultural practice in motion*. Philadelphia, PA: Temple University Press.

McLeod, K. (2003). Space oddities: Aliens, futurism and meaning in popular music. *Popular Music, 22*(3), 337-355.

Miller, D., & Slater, D. (2000). *The Internet: An ethnographic approach*. Oxford, UK: Berg.

Obrist, H.-U. (2005). Allora and Calzadilla: Talk about three pieces in Vieques. *Artforum, 43*(7), 205.

Ogborn, M. (2007). *Indian ink: Script and print in the making of the English East India Company*. Chicago, IL: University of Chicago Press.

Palan, R. (1998). Trying to have your cake and eating it: How and why the state system has created offshore. *International Studies Quarterly, 42*(4), 625-643.

Paravisini-Gebert, L., & Romero-Cesareo, I. (2011). Repeating islands: Caribbean cultures in cyberspace. *SX Salon.* Retrieved from http://smallaxe.net/wordpress3/discussions/2011/02/27/repeating-islands-caribbean-cultures-in-cyberspace/#more-65

Paton, W. A. (1888). *Down the islands: A voyage to the Caribbees.* London, UK: Kegan Paul, Trench.

Prest, J. (1981). *The Garden of Eden: The botanic garden and the re-creation of paradise.* New Haven, CT: Yale University Press.

Saldanha, A. (2011). The itineraries of geography: Jan Huygen van Linschoten's *Itinerario* and Dutch expeditions to the Indian Ocean, 1594-1602. *Annals of the Association of American Geographers, 101*(1), 149-177.

Sheller, M. (2003). *Consuming the Caribbean: From Arawaks to zombies.* London, UK: Routledge.

Sheller, M. (2004a). Demobilising and remobilising the Caribbean. In M. Sheller & J. Urry (Eds.), *Tourism mobilities: Places to play, places in play* (pp. 13-21). London, UK: Routledge.

Sheller, M. (2004b). Natural hedonism: The invention of Caribbean islands as tropical playgrounds. In D. Duval (Ed.), *Tourism in the Caribbean: Trends, development, prospects* (pp. 23-38). London, UK: Routledge.

Sheller, M. (2007a). Retouching the "untouched island": Post-military tourism in Vieques, Puerto Rico. *Téoros, 26*(1), 21-28.

Sheller, M. (2007b). Virtual islands: Mobilities, connectivity, and the new Caribbean spatialities. *Small Axe: A Caribbean Journal of Criticism, 24*(2), 16-33.

Sheller, M. (2009a). Infrastructures of the imagined island: Software, mobilities, and the architecture of Caribbean paradise. *Environment and Planning A, 41*(6), 1386-1403.

Sheller, M. (2009b). The new Caribbean complexity: Mobility systems, tourism and spatial rescaling. *Singapore Journal of Tropical Geography, 30*(2), 189-203.

Sheller, M. (2010). Air mobilities on the U.S.-Caribbean border: Open skies and closed gates. *Communication Review, 13*(4), 269-288.

Sherman, S. (1996). *Finance and fictionality in the early eighteenth century: Accounting for Defoe.* Cambridge, UK: Cambridge University Press.

Trouillot, M.-R. (2003). Anthropology and the savage slot: The poetics and politics of otherness. In M.-R. Trouillot, *Global transformations: Anthropology and the modern world* (pp. 7-28). New York, NY: Palgrave Macmillan.

Vannini, P., Baldacchino, G., Guay, L., Royle, S., & Steinberg, P. (2009). Recontinentalizing Canada: Arctic ice's liquid modernity and the imagining of a Canadian archipelago. *Island Studies Journal, 4*(2), 121-138.

Walcott, D. (2000, June 15). A frosty fragrance. *The New York Review*, p. 61.

Warf, B. (2005). Telecommunications and the changing geographies of knowledge transmission in the late 20th century. In N. Fyfe & J. Kenny (Eds.), *The urban geography reader* (pp. 353-363). London, UK: Routledge.

Wennerlind, C. (2011). *Casualties of credit: The English financial revolution, 1620-1720.* Cambridge, MA: Harvard University Press.

3

Knowing Flows:
How Migration Research Meets Mobilities Through Digital Technology

Rob Shields

This chapter briefly considers how digital technologies for computing, storage, and visualization of migration data have had major effects on the way migration as the movement of populations is increasingly understood to be consonant with multiple mobilities associated with individual bodies, their traits, capacities, and relations—with major implications for governance and for the relation of States to these populations. The approach here is to sketch the development of migration research historically. Theories such as the "gravitation model" of movement to cities as well as empirical challenges will be read through the lens of mobility studies as a history of ideas of migration, both internally and across national borders. These are also technologies, but cultural means of grasping the world as a space in which mobility is understood to be possible and desirable in specific ways, and in which mobilities are appropriate and proper to certain objects and bodies and themselves offer a type spatial technology which provides a solution to certain sorts of problems. Mobilities, alongside all technologies, have both a historicity and a spatiality. The current policy-research nexus will be considered in the form of one example, the international Metropolis Project.

 I begin by introducing the emergence and development of migration research and discuss those interconnections to mobilities studies that are increasingly obvious but have tended to not be widely discussed. I argue that this neglect has been in part due to the empirical (read: quantitative) and policy focus of migration research and the domination of the field by efforts to collect, quantify, and digitally analyze migration data. Originally, migration was a historical preoccupation of colonial North America in the face of distance and of a relative lack of known reference points. Later, social sciences and settler societies faced the challenge of establishing national identities, normative order, and spatialization in the context of waves of migrants arriving in its major gateway cities. This finally brought to light the technological trends I examine.

Migration

It is striking that the last decade of mobilities research in journals such as *Mobilities, Theory, Culture and Society* and *Space and Culture,* has not more strongly addressed migration and migration policy and research to date, given the centrality of population movements to worldwide political tensions, economic development, and policy priorities since the late twentieth century.[1] On the side of migration research, technical empirical reports and academic papers share a substantive interest with mobilities research, but this interest is neither acknowledged nor well developed theoretically, nor do researchers regularly entertain a dialogue between these two distinct theoretical traditions.

Ongoing movements from rural to urban centers in countries such as China and India are now said to constitute the largest movements of humans in history. These trends, amongst others, remind us that, historically, human populations are both settled and mobile and that histories of modernization show how economic development has turned on the availability of masses of wage laborers concentrated in cities. Not only urbanization,[2] but refugee flows[3] accompanying wars, famines, and impoverishment are major forces for rearranging the social organization of the countries migrants flee from and those in which they arrive. Immigration continues to be a major policy focus for nations globally and surely, besides transportation and its related technologies and infrastructures, it now represents the focus of the bulk of human intelligence applied to this topic (Diken & Albertsen, 2001).

Although statistical methods had to wait for the arrival of computing power, the technopolitics of immigration has long been a facet of North American nation-building. The offer of free land and of a one-way passage to a "New World" was promoted through handbills, newspaper advertisements, and word of mouth in economically disadvantaged areas of Europe, in particular. Immigration was a feature of the biopolitical organization of national territory through rural settlement and economic development by way of replacing hunting and herding with sedentary cultivation, particularly on the Prairies but also in the Midwest. For example, exploration and surveying were conducted with a view to marking out land for Caucasian settlers. Migration as promoted in North America produced differential mobiltiies. That is, the mobilities of raced and ethnically privileged "white" bodies prevailed over the mobilities and freedoms of Indigenous, Asian and African immigrant bodies. On these lines more than on others, people experienced different histories of mobility. In some cases mobility reached only a certain limit, in others it was forced, and others freedom of mobility was expropriated. It would be hard to prove otherwise from the historical record. But how have the idea of and discourses on mobility figured in historical research on migration?

On a cultural level, expeditions furnished an early source of national mythology for Canada and the United States in particular, two countries that eulogized voyages of exploration by fur traders and surveyors, such as the 1804-1806 Lewis and Clark Expedition. Greenwood and Hunt (2003) suggested that nineteenth century European and American urbanization driven by rural-to-urban migration, plus overseas immigrants to North American cities, spurred an interest in migration research. This was largely reactive research, based on observation and in some cases based on firsthand, ethnographic research in the new ghettos and tenement areas of the inner cities (for an introduction to the large body of literature on the ghetto street corner, see Park, 1925). The adjustments that foreign immigrants who did not speak the dominant language were forced to make, and the movement of African Americans and Whites from the Southern States to northern cities, posed problems of dislocation and raised questions of social identity for governments and cities (Greenwood & Hunt, 2003).

The Depression of the 1930s further raised research and policy problems, as urban unemployment was swelled by rural-to-urban migrants (Greenwood & Hunt, 2003; Thomas, 1938). The ironic link between occupancy, race and class, and between mobility and immobility in these sites, however, is little examined theoretically even though successive generations of anthropologists tarried at street corners that were some of the iconic reference points of subsequent waves of American social science research (Hannerz, 1969, to name only one member of the later generation).

From point of view of theories of urban ecology, "succession" and "displacement" by wave after wave of newcomers were probably the most important time-space features of a "space of representations" (Lefebvre, 1992) that came to frame or to "spatialize" how the relationship between populations, places, and mobility were understood and lived as constellations of problems, explanations and likely solutions. Under the rubric of the time, social problems of the late nineteenth- and early twentieth-century settlement era were but temporary, because the populations were in transition to full integration and moving on to other places. Those who became permanent residents of the inner cities were the focus of later social science and urban planning research on race and class in the North American city (Marcuse et al., 2009; Mendez, 2009; Sugrue, 2003; cf. Olalquiaga, 1992; Ricourt & Danta, 2003; Villa, 2000). Often they appeared to be populations that needed to be removed, by way of clearing inner-city neighbourhoods (Ricourt & Danta, 2003).

This late nineteenth-century space of representations frames the problem perception, research interests and choice of technologies of representation, such as mapping, censuses, physical planning and slum clearance. These social technologies in turn became lenses through which immigration as mobility was perceived and known. One conception of cities emphasized mobility flows of migrants first into city centre reception areas (ghettos) and then outward. Concentric rings of increasing levels of social integration and wealth reflected this. Poverty in inner-city ghettos was either a natural condition of impoverished immigrants or of stalled social mobility reflected in geographical stasis. Later research suggested that disadvantages in inner-city and inner-suburban neighborhoods were related to economic conditions, not the movement of the poor or of marginal newcomers (Cooke, 2010). Indeed, migrants are not only impoverished entrants at the bottom rungs of the social ladder, but increasingly they are affluent professionals attracted to top-tier global cities (Buss, 2002; Tastsoglou & Miedema, 2003). They are also attracted to ongoing mobility practices which dissociate their workplaces from family abodes, extended families, and their preferred leisure and retirement destinations from their countries of citizenship (Cervantes-Rodriguez, Grosfoguel, & Mielants, 2009; Janoschka, 2009; Smith & Guarnizo, 2009; Vertovec, 2004).

First beginning with 1920s and 1930s census data, demographers and sociologists dominated North American migration research with a stress on arrival and assimilation. Geographers soon noted that these data showed,

> a remarkable mobility. Fewer than half of the families of the United States are bound to a locality by the ties of home ownership, and the automobile has destroyed all respect for distance. But since the beginning of the depression, migration has assumed the form of bewildered aimlessness. (Thornthwaite, 1934, p. 3)

There is a consistent rhetoric of migration as both an opportunity and a problem in these social science and policy discourses. There is a sense of policy reacting to trends, and of flows such as immigration or movement to cities as being out of the control of the State. While flows may be channelled and are reversible, there is the sense of waves which are not only a rhythm or cycle of migration driven by economic and environmental factors (e.g., the poor crops of the U.S. Midwestern "Dust Bowl" of the 1930s), but an irresistible force. Invesment in scientific knowledge projects and technoliges was spurred by the collapse of a space of representations in which a "mobility practice" such as migration was a hard but proven route to betterment. Moving solved problems.

In relation to mobility and flow, we can read Greenwood and Hunt's (2003) short history of migration research as a history of ideas. The broader social science interest was policy-driven by questions such as those posed by Goodrich and colleagues in their book, *Migration and Economic Opportunity*: "If men and women are out of work, can they hope to gain employment by migration? If their living is precarious, would they be more secure in other locations?" (Goodrich et al., 1936, n.p.). Thus, with the rise of the welfare state, migration studies became most obviously yoked to preoccupations at the scale of the territorial nation state. Mobilities became a problem not a sure solution. The scales at which migration studies is focused has been driven not only by the availability of national censuses and research funding but by a preferential optic that blurs the foreign and origin states, however powerful these may be in continuing to shape peoples' Diasporic communities, identities and politics—especially through religion and language. Would the unemployed be more likely to find a job if they migrated? And, should working people move to find better paying jobs, or stay where they are? Today we continue to find these kinds of questions raised in reaction to local disadvantage and disaster. For example, after Hurricane Katrina devastated some parts of New Orleans in 2005, economists advocated that the city should be abandoned and the inhabitants moved elsewhere, neglecting the fact that 70% of the city was never flooded (see Steinberg & Shields, 2008).

By contrast to this approach driven by the interpretation of demographic data, early British geographers such as Ravenstein, had proposed a scientific law of migration whereby people tended to gravitate to larger centers (Ravenstein, 1885). Geographically attuned economists such as Marshall emphasized the attraction of the city as a form of social selection: "large towns and especially London absorb the very best blood from all the rest of England; the most enterprising, the most highly gifted, those with the highest physique and strongest characters go there to find scope for their abilities" (Marshall, 1948, p. 199). "Gravity models" such as this continued to play an important role in discussions of causal economic drivers of movement in both the US and the UK. These clearly suggest "spatializations" in which more than natural forces inexorably attract immigrants. However, there was a shift in the US. This was due, firstly, to a turnaround in the experience of the 1930s Depression, secondly, to the rapid development of new technologies that displaced the traditional tempo of lifetimes of learned craft, skill, and thirdly, changing local population levels driven by biological factors such as the birthrate, life expectancy, and mortality. Thomas (1938) commented on the American experience of the 1930s, stressing migration differentials rather than absolute "laws":

> Migration may be viewed as a sort of goal-seeking, in which economic motives are involved (the migrant attempts, or hopes, to better his economic condition), but in which hedonistic motives also play an important role (the migrant attempts to escape the monotony of a limited environment, to exercise his right of choosing a new environment, and hopes, through migration, to attain and enjoy a richer environment). It also suggests the possibility that some migrations are, or appear to be, unmotivated, random movements.

> Defining differentials in connection with these problems requires much more than meticulous application of statistical methods. Data obtained as administrative by-products are, in the main, unsuitable for throwing much light on these problems. The behavior of the migrants must be observed before and after migration; the migrants' "own stories" must be obtained; the environmental setting and the conditions of life in the communities of origin and destination must be described. This does not mean that statistical methods are inapplicable; on the contrary, they will be essential in the initial process of sampling and in the final evaluation of the observed differentials. (pp. 141-142)

This "before and after" statistical approach, filled in by qualitative information, set the standard for quantitative social science approaches which focused on stable residence with movement as an irregular, interstitial moment. While migrants began to be surveyed directly, the tendency for people to migrate was attributed to places and regions on a state by state basis, fixing mobility to static locations (Kuznets & Thomas, 1960). The sense is that geographic space is static, a fixed game board on which pieces move without changing the board. All of history indicates however that immigrants shape cities and settlers transform landscapes. Migrants weave ties between places and technologies shrink perceptions of distance. Space is also changeable, a fluid and thus mobile spatialization (Shields, 2006; in press). Even geographers observing the migrations of the 1930s, commented that "since the movement is abnormal in most respects, it is inconceivable that it will continue" (Thornthwaite, 1934, p. 18).

Spatializations are cultural technologies that differentiate and structure mobilities by "torquing" the relationships between locations to render some more attractive or with a character they would otherwise not have. Interrupted by World War II, a later generation of researchers in the 1960s translated the distance and population variables of the gravity model into regression formulas to test hypotheses regarding the most important factors behind migration. This meant a move from approaching mobilities occurring in a *topology* (Shields, in press) warped by a disequilibrium of wages, opportunities, and amenities between locations, to a consideration of employment change and individuals' households' life-cycle events and the consequences of migration on places and regions. Central in this trend was the increasing ability of computers to process large amounts of data. This allowed surveys to be extended from simple variables such as origin and destination to include individual and household characteristics.

Metropolis: Knowledge Technologies

Recent social science migration research in the OECD countries is perhaps most famously crystallized in the Metropolis project: an international network on migration funded by both academic granting councils and government departments which draws together public servants and policy researchers. On the one hand, Metropolis represents a form of state capture of research. On the other hand, in its avowed policy relevance and multidisciplinarity, Metropolis represents the turn to evidence-based policymaking as well as the integration of the demographic profiling and mapping of earlier decades onto a conscious policy concern with managed, sociocultural diversity.

Over the years, the consortium of researchers in the Metropolis project has documented,

> The specific strategies that immigrant groups employ to effect successful integration within urban structures and systems, the processes by which these strategies are pursued and modified, and the outcomes of these processes.... The goal is to better understand the processes by which immigrants become Canadians. (Prairie Metropolis Centre, 2010, n.p.)

Integration, urban issues and the restructuring of identity are heavily coded into the mandate of the research. Regardless of social and political stances toward assimilation, the international, comparative research under the Metropolis project has considered integration as a core focus. The ongoing, continued mobilities of migrants whose gaze looks selectively to countries of origin is a puzzle to be examined and, in some cases, resolved into static dwelling and a settled identity. There is a tendency to organize the narrative of migration mobilities through a history of places of arrival, such as "North American city." Integration also finds a mirror in xeno-racisms, policy barriers to success, resistance toward and rejection of new arrivals by established residents.

Early research explored the breadth of this mandate but most efforts have emphasized the aspects of mobility and migration that matter most to national governments. This develops themes established in the early 1960s that approach migration as the national acquisition of human capital and skilled labor, often understood through the lenses of certification and qualifications and the relative amounts and directions of flows (e.g., Sjaastad's (1962) discussion of backflows).

Despite the international research scope, the national mandates of Metropolis groups—notably in countries such as Canada—have a geographical focus on borders and on policies that solicit, admit, or exclude immigrants. The research is thus also comparative. However, geographical context becomes easily abstracted: the focus is on differentiating the population, on citizenship as an ideological and behavioral practice within the context of the State, the official public sphere of law, and political participation—in contrast to the micro and sub-public sphere of comportment and family containment within the single-family dwelling unit.

As contemporary examples of migration research, several other key problematic issues are apparent:

1. Most commonly, there is an oversimplification of migration which replaces mobilities with a simple, one-way movement of arrival, and complex identities with a simple national model of the multicultural citizen—that is, a bicultural citizen with a hyphenated identity and possibly a dual citizenship.

2. Cultural imaginaries have too long been off the agenda because of a lack of theoretical and ontological frameworks to distinguish social categories such as community (virtual ideal-real entities) from mere concepts or fictions (ideal possibilities such as abstractions and ideas; see Shields, 2003).

3. There is a disconnect between a materially specific, urban context and the abstract space of population tables.

4. There is poor integration between the official public life of cities and the State, and a suppressed sphere of everyday immigrant life that has been pushed away from the public sphere and uneasily accommodated in public spaces (from squares, to playgrounds, to parks and recreational facilities).

5. A displacement of the sphere of political struggle to a sub-public sphere.

In North America, historically, immigrants have tended to settle in large metropolitan areas, despite attempts by state authorities to disperse them. Cities are the locus of migration but do not find recognition in the relevant legal and constitutional frameworks. In Canada, for instance, migration is managed on a provincial basis, with limited consideration of immigrants' secondary moves to the most economically or ethnically appealing cities. Sustaining cities requires that migration be understood as a driver producing urban space and forms, often in suburban, middle-zone, and intra-urban areas. Along with country-to-city and regional flows, migration is thus integral to urbanization and suburbanization. Despite the cultural fluidity and changing collective identity that transnational and rural migrants bring to urban areas, the combination of rapid population growth, a demand for services, and pressures on infrastructure and provision has been argued to contribute to metropolitan poverty, segregation, and the rise of new discriminations and social stratifications—outcomes that work against the state goals of integration to full citizenship, and the preservation of social order.

Immigration policy tends to focus on a national scale, though urban localities are where migration is most tangibly felt and where difference is lived. Data at this local scale is usually not available. Nonetheless, analysis in many countries tends to focus in particular on quantitative profiles. However, migrant social life and the negotiation of ethnic community services and identity are key aspects of migration as an ongoing engagement with a diasporic lifestyle in the Americas (Seo & Kang, 2010). The actual processes of immigrant place-making get still less attention—especially in the context of suburban neighborhoods. Places and neighborhoods have a role in hosting and "allowing" migrant identities and diasporic communities to take tangible shape and be performatively actualized in both seasonal displays and the everyday routines. These make culture present, tangible, and a lived force in urban environments. Community institutions such as Vietnamese pagodas are sites of presence, and neighborhoods are spaces in which virtualities and abstractions such as migration flows become actually felt and translated into environments of predictable interactions, risk, and opportunity. The built environment offers more than semiotics and shelter. By signaling cultural identification and presence, it establishes an ethos where cross-cultural encounters and interactions are anticipated. This affective ecology from social interaction normalizes cultural difference.

Traditionally, enclaves have been understood as "ghettos" and forms of temporary residence or refuge for the new, poor migrants excluded by cultural, politico-economic, and language barriers. However, especially around the Pacific Rim, contemporary migrants are resource rich in a variety of ways, not the least of which is as representatives of transnational capital. Ethnic enclaves may thus also be economic gateways. The possibility of travel, and media flows of cultural, political, and business information encourage new forms of migration (Chambers, 1987) and new orders of mobilities. This leads to a more dynamic urban population composed of people from a shifting mosaic of backgrounds in complex and fluid spatial concentrations. Fixed neighborhood identities with less heterogeneous populations and more stable tenure become hallmarks of both the ghettoized disadvantaged and of the economically and socially better off: a clearly geographical pattern of enclaves; arrested mobilities in the former case and "detourned" and excluded mobilities in the latter.

What is most striking about this brief history of quantitative migration studies is its youth. Despite its entrenched character in policy and in social science departments, the time span of effective analysis of large data sets and of multiple regressions has been very short, dating only from the 1970s. Shifts to more dynamic approaches suggest that migration studies are undergoing a parallel development to mobilities studies, despite the apparent lack of conversation between the two literatures through 2010. Lindley cites Castles to note that,

> in exploring how a shifting structural environment interacts with individuals' and families' unique configurations of capabilities and resources to produce migration, we avoid designating the displaced as a homogenous mass *in*flux, rather than illuminating variability and agency in people's responses...by firmly situating their analysis in relation to the wider global political economy in which these micro-level realities unfold. (Castles, 2003, p. 17, as cited in Lindley, 2010, pp. 17-18)

For example, in the case of refugee streams from war, violence, and persecution, micro-level, fine-grained case studies of conflict and mobility can make it clear that these are "not the result of a string of connected emergencies" (Castles, 2003, p. 17, as cited in Lindley, 2010, p. 18), but are driven by structural forces at the global level. Similarly these migration flows are now well explained by resort to determining variables impacting populations, but require an attention to household histories as a space of distance, difference and mobility, as well as cultural perceptions of the world.

Information and Computing Technologies

The role of technology in mobility studies and in migration research generally foregrounds the emergence of long-distance transportation systems and communications networks, from media that publicized destinations, to the postal system and freight distribution networks that allowed people to stay in touch and exchange goods. However, for studies of migration in the social sciences, the development of information and computing technologies for data gathering, quantification, and analyses has been central. These new tools in turn have an effect as technologies of individuation, surveillance, and governance.

While census taking had been used since the time of the early Chinese and Egyptians as a tool for both taxation and, later, military conscription (Statistics Canada, 2011). Migration at state borders can be tracked, but flows in and out of cities is much more difficult to estimate, hence the tendency of cities to conduct more frequent municipal censuses for tax and service-provision purposes. However, the ability to break down overall numbers by needs and by particular groups has not derived solely from the development of statistical methods by Fermat and Pascal in the 1650s (Hacking, 2006), or on de Quetelet's conceptions of the average as a means of understanding social phenomena in the early 1800s. The ideas of inference, regression (Peirce & Jastrow, 1885), and random sampling championed by C. S. Peirce in the late 1800s (Peirce, 1931-35, 1958, v.7, pp. 138-157) have also been especially important.

All of these mathematical techniques depended on organizational, identification, and recording techniques such as the humble paper form to be filled out, and the urban ward postal system. Original information gathering was done by census takers who wrote down household information in pencil, which was then transcribed to bound volumes resulting in numerous misspellings. At borders also, the spellings of foreigners' names were often romanized and/or anglicized, willingly or not. For census taking, from 1950 onward in the US, census forms were mailed to every address on record and in 1970 it was made illegal to fail to complete and return these forms, with canvassers used to verify a random sample. Only in the 1970s was computer technology introduced to consolidate these two sets of forms. The raw data changed from paper-based to magnetically encoded and optical media.

Informational technology is central in more than the development of statistical analyses and projections. The computer power necessary to conduct complex regressions only dates from the mid to late 1970s. Even at this time, the steps in analyses were limited by the capacity of researchers to work with paper punch cards, each one of which often carried a single piece of code or memory address. For university-based researchers, the ability to interrogate large data sets dates from the late 1980s, but more properly, from the early 1990s, with the arrival of personal computers with sufficient computing speed and magnetic hard drives of sufficient size to hold the contents of a large survey data matrix. The wider usability of large data sets and sophisticated software suites such as **SPSS** was then simplified by the availability of **CD-ROM** optical media introduced by the Sony and Phillips corporations in the mid-1980s.

The place-based emphasis on data gathering by household and postal address began to shift with graphical visualizations of flows between places, made possible by computer graphics and display technologies in the 1990s. This is a further example of how migration comes to be better known as "flows" and as "mobilities" rather than as single movements, like discreet hops from one point to another. The ability to visualize flows of populations, particular groups, and financial data leads to an increased understanding of the range of mobilities around population migrations, such as remittances to family remaining behind or fears of a "brain drain" of particular professions and skilled labor. Computing technologies have impacted the cultural technologies that set frameworks by which populations were imagined and policies conceived.

The ability to more finely govern migration, that is mobility itself, as opposed to migration as an outcome of movement, is also expanded by the "partability" of these flows, where data about bodies and objects can be separated from the material of the migrant body or cargo and can be processed in different ways, other places, and ahead of time, or audited at a later date (Chalfin, 2007). Health and other personal information is no longer collected only on arrival at the border but well in advance at consulates or by third party travel services from health clinics to immigration lawyers. Borders themselves are hardened into physical walls and gates, parcellizing the space of North America still further and intervening in the possibilities of actual mobility. The multiplicity of transfers may make one wonder when the time and point of departure or arrival actually takes place; more often than not bodies will arrive leaving key aspects such as their property, financial holdings and citizenship in place behind. Similar to the exile, one may experience a separation of the components of one's everyday life between places (and times). This has radical implications for the governance of migration as mobility. It aligns with service delivery to individuals rather than place-based neighborhoods or communities with social needs, in effect shifting the presuppositions on which service delivery in the welfare state was based toward new neoliberal models.

The partability of the elements of migration—data, material, capacities, relations—necessarily means that migration will more and more reference bodies and capacities per se and is in need of a broader command of the mobilities on which these bodies depend, whether in motion or at rest. Migration comes then to be understood through knowledge technologies of data gathering, recording, and analysis. This in turn changes the way migration is seen. When we witness through visualization rather than encounter and react to gathered masses, the rhetoric and understandings of mobilities seem to come to the fore. In addition, rather than being perceived as a purely urban concern, the problems of migration are understood at a scale that includes origin and destination. As these become mere waypoints in a life cycle of mobility, the scale shifts from the nation state to the global.

These comments are brief and preliminary. However, the implications of the occasion when migration and mobilities research intersects can be seen to be significant in terms well beyond the phenomenon itself. Technologies such as computing devices and information processing capacities have made immigration not only more visible as an aggregate statistic but more amenable to policy intervention. New computing technologies impact on the informational, social and cultural technologies by which immigration is understood to be a problem-solving mobility of "departures and arrivals" from static places. These were understood within a fixed spatialization via relatively fixed images such as the concentric rings of urban neighbourhoods that were indicative of success, expanding out in relation to inhabitants' status. Geographical space was a context of mobility. However, who moved and where was the subject of struggle that produced an ethnically and racially divided space of populations.

Notes

1. The exceptions to this rule are generally outside of mobilities research itself (for example, the work of Dasgupta (2008)).

2. I do not deal with the city both as a nexus and form of mobilities here. The city is more than the site of mobilities or the scene where mobilities can be observed as economic or labour market processes. The city itself is a nexus of mobilities, woven out of myriad flows. This "ontogenetic," rather than ontological conception of space (Adey, Budd, & Hubbard, 2007, p. 79) brings relations and flows to the foreground rather than treating them as fixed objects in a static geometry (see Shields, 2006).

3. I do not discuss flows per se in this chapter (see Shields, 1997). Throughout the literature, flow is advanced as the definition of mobilities. As a concept and figure, flow is a primordial, theoretical form of "mobilities." But what are its theoretical affects (Davidson, Park, & Shields, 2011)—its capacities, qualities, and potential as, for example, viscous or inertial, directional stream, and diffuse sprawl? What are the topological limits and presuppositions of such surface mobilities, and what spatializations are implicit (e.g., a stable, orientable space which serves as the context for a non-orientable mobility (Shields, in press)? Thus the assertion that "the global presupposes the metaphors of network and flow rather than that of region" (Urry, 2000, p. 32). And the literature on globalization and on mobilities has been critiqued as a chronically disembedded, anti-human "flow-speak" (Bude & Dürrschmidt, 2010).

References

Adey, P., Budd, L., & Hubbard, P. (2007). Flying lessons: Exploring the social and cultural geographies of global air travel. *Progress in Human Geography, 31*(6), 773-791.
Bude, H., & Dürrschmidt, J. (2010). What's wrong with globalization?: Contra 'flow speak'—towards an existential turn in the theory of globalization. *European Journal of Social Theory, 13*(4), 481-500.
Buss, T. F. (2002). Emerging high-growth firms and economic development policy. *Sage Urban Studies Abstracts, 30*(3), 279-411.
Castles, S. (2003). Towards a sociology of forced migration and social transformation. *Sociology, 37*(1), 13-34.
Cervantes-Rodriguez, A. M., Grosfoguel, R., & Mielants, E. (2009). *Caribbean migration to Western Europe and the United States: Essays on incorporation, identity, and citizenship*. Philadelphia, PA: Temple University Press.

Chalfin, B. (2007). Customs regimes and the materiality of global mobility: Governing the Port of Rotterdam. *American Behavioral Scientist, 50*(12), 1610-1630.

Chambers, I. (1987). Maps for the metropolis: A possible guide to the present. *Culture Studies, 1*(1), 1-21.

Cooke, T. J. (2010). Residential mobility of the poor and the growth of poverty in inner-ring suburbs. *Urban Geography, 31*(2), 179-193.

Dasgupta, S. (2008). Between the aesthetics of migration and migratory aesthetics. In G. Watson, A. Dasgupta, M. Szewczyk, & W. Bradley (Eds.), *Santhal family: Positions around an Indian sculpture* (pp. 98-105). Antwerp: Museum van Hedendaagse Kunst Antwerpen.

Davidson, T., Park, O., & Shields, R. (Eds.). (2011). *Ecologies of affect: Placing nostalgia, desire, and hope.* Waterloo, Ontario, Canada: Wilfred Laurier University Press.

Diken, B., & Albertsen, N. (2001). Mobility, justification, and the city. *Nordisk Arkitekturforskning–Nordic Journal of Architectural Research, 14*(1), 13-24.

Goodrich, C., Allin, B. W., Thornthwaite, C. W., Brunck, H. K., Tryon, F. G., Creamer, D. B., et al. (1936). *Migration and economic opportunity: The report of the study of population redistribution.* Philadelphia, PA: University of Pennsylvania Press.

Greenwood, M. J., & Hunt, G. L. (2003). The early history of migration research. *International Regional Science Review, 26*(1), 3-37.

Hacking, I. (2006). *The emergence of probability: A philosophical study of early ideas about probability, induction and statistical inference.* Cambridge, UK: Cambridge University Press.

Hannerz, U. (1969). *Soulside; inquiries into ghetto culture and community.* New York, NY: Columbia University Press.

Janoschka, M. (2009). The contested spaces of lifestyle mobilities: Regime analysis as a tool to study political claims in Latin American retirement destinations. *Erde, 140*(3), 251-274.

Kuznets, S., & Thomas, D. S. (Eds.). (1960). *Population redistribution and economic growth, United States, 1870-1950, Vol. II: Analyses of economic change.* Philadelphia, PA: American Philosophical Society.

Lindley, A. (2010). Leaving Mogadishu: Towards a sociology of conflict-related mobility. *Journal of Refugee Studies, 23*(1), 2-22.

Marcuse, P., Connolly, J., Novy, J., Olivo, I., Potter, C., & Steil, J. (Eds.). (2009). *Searching for the just city: Debates in urban theory and practice.* London, UK: Routledge.

Marshall, A. (1948). *Principles of economics: An introductory volume* (8th ed.). New York, NY: Macmillan.

Mendez, P. (2009). *Immigrant residential geographies and the 'spatial assimilation' debate in Canada, 1997-2006.* Vancouver, British Columbia, Canada: University of British Columbia Press.

Olalquiaga, C. (1992). *Megalopolis: Contemporary cultural sensibilities.* Minneapolis, MN: University of Minnesota Press.

Park, R. E. (1925). The city: Suggestions for the investigation of human behavior in the urban environment. In R. Park & R. Burgess (Eds.), *The city* (pp. 1-46). Chicago, IL: University of Chicago Press.

Peirce, C. S. (1931-35, 1958). *Collected papers of Charles Sanders Peirce.* Cambridge, MA: Harvard University Press.

Peirce, C. S., & Jastrow, J. (1885). On small differences in sensation. *Memoirs of the National Academy of Sciences, 3,* 73-83.

Prairie Metropolis Centre. (2010). *Metropolis Project annual report 2009-10.* Edmonton, Alberta, Canada: Author.

Ravenstein, E. G. (1885). The laws of migration. *Journal of the Statistical Society of London, 48*(2), 167-235.

Ricourt, M., & Danta, R. (2003). *Hispanas de Queens: Latino panethnicity in a New York City neighborhood.* Ithaca, NY: Cornell University Press.

Seo, J., & Kang, M.-J. (2010, March). *The subjective construction of urban space by migrants.* Paper presented at the Urban Affairs Association 40th Annual Meeting, Honolulu, HI.

Shields, R. (1997). Flow. *Space and Culture—Theme Issue on Flow, 1*(1), 1-5.

Shields, R. (2003). *The Virtual.* London, UK: Routledge.

Shields, R. (2006). Knowing space. *Theory, Culture & Society, 23*(2-3), 147-149.

Shields, R. (in press). *Topologies of space.* London, UK: Sage.

Sjaastad, L. (1962). The costs and returns of human migration. *The Journal of Political Economy, 70*(5), 80-93.

Smith, M. P., & Guarnizo, L. E. (2009). Global mobility, shifting borders and urban citizenship. *Tijdschrift voor Economische en Sociale Geografie, 100*(5), 610-622.

Statistics Canada, (Ed.). (2011). *The Canadian encyclopedia.* Toronto, Ontario, Canada: Historica-Dominion Institute.

Steinberg, P. E., & Shields, R. (Eds.). (2008). *What is a city?: Rethinking the urban after Hurricane Katrina.* Athens, GA: University of Georgia Press.

Sugrue, T. J. (2003). Revisiting the second ghetto. *Journal of Urban History, 29*(3), 281-290.

Tastsoglou, E., & Miedema, B. (2003). Immigrant women and community development in the Canadian maritimes: Outsiders within? *Canadian Journal of Sociology-Cahiers canadiens de Sociologie, 28*(2), 203-234.

Thomas, D. S. (1938). *Research memorandum on migration differentials.* New York, NY: Social Science Research Council.

Thornthwaite, C. W. (1934). *Internal migration in the United States.* Philadelphia, PA: University of Pennsylvania Press.

Troper, H. (2000). *History of immigration to Toronto since the Second World War: From Toronto "the good" to Toronto "the World in a City."* Toronto, Ontario, Canada: CERIS.

Urry, J. (2000). *Sociology beyond societies: Mobilities for the twenty-first century.* New York, NY: Routledge.

Vertovec, S. (2004). Migrant transnationalism and modes of transformation. *International Migration Review, 38*(3), 970-1001.

Villa, R. (2000). *Barrio-logos: Space and place in urban Chicano literature and culture.* Austin, TX: University of Texas Press.

4

If Only It Could Speak:
Narrative Explorations of Mobility and Place in Seattle

Ole B. Jensen

This chapter imaginatively engages with the ways in which place is understood by telling a story about urban transformation in Seattle. Empirically this is explored by "giving voice" to the reconstruction of State Route 99 (SR99) in downtown Seattle (a project estimated to cost in excess of US$3 billion). According to the 2010 census there are 608,000 residents in the Seattle city area and some 3.4 million inhabitants in the metropolitan area. SR99 runs 50 miles from Fife in the south to Everett in the north and in parts in parallel with the Interstate 5, however closer to the coast. The SR99 cleaves its way on the north-south trajectory through the city via raised ramps and elevated sections and carries about 110,000 vehicles a day.

One of these sections, the Alaskan Way Viaduct, was built on old seawall structures and created a buffer between the city and the sea at Elliott Bay. The Alaskan Way Viaduct, colloquially known as the "Seawall," was opened on April 4, 1953 and extensively damaged by an earthquake in 2001. The subsequent debate concerning plans for its reconstruction serves as a lens for understanding the complex relationship that exists between human and non-human elements and illustrates how infrastructures and mobility systems are simultaneously both material and cultural artifacts that need to be understood very differently from the utilitarian and instrumental perception that guides much contemporary urban planning and design.

Inspired by Bruno Latour's (1996) story of the aborted *Aramis* light rail project in Paris and Phillip Vannini's (2008) account of the sinking of the *Queen of the North* ferry in British Columbia, this chapter embarks on a similar thought experiment of imagining how a redesign of a large urban infrastructure project like the Seawall in Seattle would look "*if only it could speak.*" This approach highlights that behind any physical development project there are multiple voices which belong to different stakeholders and institutions. In order to capture the approach unfolding in this chapter I begin by quoting part of Latour's introduction:

> In this book, a young engineer is describing his research project and his sociotechnological initiation. His professor offers a running commentary. The (invisible) author

adds verbatim accounts of real-life interviews along with genuine documents....Mysterious voices also chime in and, drawing from time to time on the privileges of prosopopoeia[1], allow Aramis to speak. These discursive modes have to be kept separate if the scientification[2] is to be maintained; they are distinguished by typography. The text composed in this way offers as a whole, I hope, both a little more and a little less than a story. (1996, p. x)

By "allow[ing] Aramis to speak" Latour was able to articulate different perspectives or voices that are involved with the project.

In the present story four different voices have been identified; that of the author, that of academic theorists, the empirical voices of various stakeholders and institutions in the field, and finally, the "voice" of the Seawall itself. Again Latour serves as an inspiration, as he spoke of the complex relationship between humans and our "inferior inanimate brothers." He wrote: "our collective is woven together out of speaking subjects, perhaps, but subjects to which poor objects, our inferior brothers, are attached at all points. By opening up to include objects, the social bond would become less mysterious" (Latour, 1996, p. viii). This chapter therefore aims to give voice to one "poor object," the Seawall, and explore the extent to which this approach may enrich our understanding of mobility and place. I may phrase it so that the artifact in question assembles multiple voices from the field constituting "place." In this chapter I draw on two theoretical perspectives in order to better examine the debates surrounding the reconstruction of Seattle's Seawall: theories of narratives and their relations to place, and theories of urban networks, socio-technical systems, and mobility.

In the 1980s, MIT Professor, Sherry Turkle, showed how computers should be understood as "evocative objects" onto which humans invest feelings (Turkle, 1984). However, emotional bonding between human and artifact extends beyond the computer and may include all sorts of other machine assemblages such as cars, trains, weapons, buildings, and ships (see Ingersoll, 2006). The latter is powerfully illustrated by Vannini's (2008) study of the sinking of *MV Queen of the North* in British Columbia on March 22, 2006. This was a story about how the wrecking of a ferry unfolded dramatic and emotive relations between a community and the material artifact. In the words of Vannini (2008, p. 156): "Funny how you can miss somebody you never met, how you can grieve the loss of a machine. Funny how she doesn't feel like a piece of metal any more."

My argument is that the Seawall may also be thought of as an "evocative object." Indeed, it may be understood as being more than an isolated artifact of tarmac, steel, and stone because it involves users and stakeholders in multiple and complex ways. Vannini spoke of "technoculture" as an expression of what people do together with things (Vannini, 2008, p. 156). Vannini's approach resonates with the way the Seawall might be understood as a complex technological phenomenon. Physical infrastructure may be understood as a relational nexus of "hardware" (e.g., asphalt, concrete, columns, traffic lights) and "software" (e.g., demand management strategies, traffic codes, and road safety campaigns). It is this assemblage of material and immaterial, human and non-human that needs to be understood in relation to the flow and friction it affords. To demonstrate this point and related arguments, this introduction leads into a brief theoretical discussion of notions of narrative and place, and the idea of place as relational and mobility-defined. This is followed by an exploration of some of the many voices involved with the Seawall's reconstruction. The final two sections offer an approach to the study of the Seawall development by allowing it to "speak" and provide final comments by way of a conclusion.

Narratives, Representation and the Politics of Place

The idea that places and sites are laying around awaiting human intervention is naïve. Indeed, as planning scholar Robert Beauregard (2005, p. 54) argued, "places are never emptied" as planners and designers substitute a professional narrative for a multitude of shared histories, collective memories, and personal experiences. Narratives abound in all fields of human activity and our species could accurately be described as storytelling animals. All our stories are situated in time and space and possess certain material requisites (Scollon & Scollon, 2003).

Since the days of Aristotle (and probably earlier) the storytelling animal has cultivated particular ways of telling and organizing narratives. Some of the elementary building blocks are the way events are structured in their relation to time and place (Flyvbjerg, 1998, p. 8). However, as Finnegan rightly claimed, "a mere listing of past events with no connecting thread does not make a story. We need something more than just temporal sequence, something to give it an intelligible plot" (Finnegan, 1998, p. 10). The notion of "plot" is key and should be understood as "the basic means by which specific events, otherwise represented as lists or chronicles, are brought into one meaningful whole" (Polkinghorne, as cited in Czarniawska, 2004, p. 7).

Narratives are not detached from the material world. Rather they always make a geographical claim (Eckstein & Throgmorton, 2003, p. 6). In other words, narrative and space need to be understood as part of the same assemblage fusing a "representational logic of urban intervention" with materiality and artifacts (Jensen, 2007, p. 218). The key is to understand the way narratives connect or disconnect to the place dimension. Moreover, what makes the whole story come alive and touch the deepest level of human emotion is the fact that there is an element of "drama" to most of our engagements with other humans and material artifacts. Vannini's (2008) description of a shipwreck is a drama in which the technological artifact plays a key role. In the same way, the damage caused to the Seawall during the earthquake is a drama triggering reactions and unfolding events. Seen this way the drama provides the narrative with a crucial before and after component. Even though this may not be clear-cut, as decisions made before an event may influence the perception of an event and thus contribute to its subsequent interpretation, most Seattleites talk about before and after the Seawall collapse. Some agents clearly use the event to articulate a "new beginning" and, as such, the drama is inscribed into the sociomaterial changes that are interpreted and retold with a particular (political) intention. As mentioned in the beginning of this chapter I shall add an additional theoretical frame for interpreting the case, namely that of a relational and mobility-defined understanding of place. It is to this second perspective that I now turn.

Place as a Relational and Mobility-Defined Assemblage

In the urban studies literature there is a long-standing debate about the relationship between the static form and morphology of the city versus all its fluid elements. In this chapter I will not go into the debate between "sedentary" and "nomadic" conceptions of cities and place in much detail (see, for example, Cresswell, 2006; Graham & Marvin, 2001; Jensen, 2009; Kolb, 2008). However, my point of departure needs to be specified. My basic analytical viewpoint is one of acknowledging how all sites, places, buildings, and cities are what they are as a consequence of the extent to which they afford, encourage, and host networked flows of people, goods, capital, and information. In the words of Dovey (2010): "Places of becoming are constructed and sustained by their connections and [there is a need to move] towards an understanding of this open sense of place" (p. xi). "[T]he task for place theory is to move from conceptions of place as stabilized being towards places as becoming" (p. 13).

If this is the point of departure, then the networked relation of sites, buildings, or cities becomes crucial. Here one should not fall into the trap of seeing footloose movement all over, but rather understand how fixity gets its meaning due to flow and vice versa (Lefebvre, 1974/1991, pp. 92-93). The city as we know it is therefore a relationally, mobility-defined node in a network of local, regional, and global connectivity. The network understanding of places and cities relate to the notion of place as "assemblage" (DeLanda, 2006; Dovey, 2010; Farias & Bender, 2010). This relates to a debate within the urban studies literature that takes the notion of assemblages as a key point to criticize the monolithic and fixed notion of cities:

> The notion of urban assemblages in the plural form offers a powerful foundation to grasp the city anew, as an object which is relentlessly being assembled at concrete sites of urban practice...as a multiplicity of processes of becoming, affixing sociotechnical networks, hybrid collectives and alternative topologies. From this perspective, the city becomes a difficult and decentered object, which cannot any more be taken for granted as a bounded object, specific context or delimited site. The city is rather an improbable ontological achievement that necessitates an elucidation. (Farias, 2010, p. 2)

In other words, the Seawall must be understood not so much as an artifact and a thing, but more as an assemblage of elements which includes everything from urban furniture, to concrete pillars, to road pavement, and mobile subjects. A State Road ceases to be a State Road if no one ever uses it for driving purposes (it may turn into other things like a party zone or a vacant dead space). Such a notion of mobility systems being complex assemblies of human and non-human elements points towards a different understanding of the citizen within the system and the way citizens are reflected in the planning and design of infrastructure. That is to say that a complex infrastructure like the Seawall becomes not only the material venue for physical mobility, but also a site of imagined future mobility. This is what takes place in the planning and policy-making process where multiple "imagined mobile subjectivities" (Richardson & Jensen, 2008, p. 218) are being narrated in stories about the future in Seattle.

From a study of European Metro systems in London, Paris, and Copenhagen, Jensen (2008) emphasized the complex interplay between artifacts, systems, and agents flowing through the system:

> trains, trails, stations, platforms, escalators, metro staff, travelers, signs, commercials, musicians, homeless, police force, tickets, ticket machines, power supplies, newspaper stands, coffee shops, customers etc. are assembled into socio-technical systems producing the lived mobility of metro travelers in London, Paris and Copenhagen. The specific assemblage within the socio-technical system is "what makes metro mobility" by means of sorting, filtering, circulating, and orchestrating mobilities. (Jensen, 2008, p. 19)

Much more could be said about the relationship between theories of narrative and place, and place as relational and mobility-defined. However, the key concept here is that complex infrastructures of contemporary urban mobility must be understood as constituting both the material/immaterial and human/non-human realms and thus we should pay particular attention to the relational and mobility-defined character of place within such infrastructures. Stories about change and transformation, as well as resistance to change (Hommels, 2005), should be seen as a complex interweaving of the material and the linguistic.

A Narrative Exploration of Mobility and Place in Seattle

Figure 1: Alaskan Way Viaduct (WSDOT)

Enough has been said from the vantage point of academic theory. Now the time has come to look at the story of Seattle's Seawall (Figure 1). Given the complexity of the story, the chapter can only highlight a few of the many voices relevant to this case. According to the official *Alaskan Way Viaduct Replacement Project History Report*:

> The Alaskan Way Viaduct section of State Route (SR) 99 has been a fixture of the downtown Seattle waterfront for over five decades....The Alaskan Way Viaduct carries about 110,000 vehicles a day and provides a convenient route to and through downtown Seattle.... However, the viaduct's days are numbered. The Nisqually earthquake and wear and tear from daily traffic have taken their toll on the facility. (http://www.wsdot.wa.gov/projects/Viaduct)

The historic reference is significant as it illustrates how prevailing technologies of mobility and their interpretations may change over time. In the 1950s, when the Seawall was first erected, it was a symbol of modern and rational engineering which was characteristic of infrastructure projects in the "Western" world and in North America in particular.

From its genesis the Seawall was the result of a complex relationship between human and non-human elements. The presence of the Seawall dominated a part of Seattle for decades and it became not only an important link for transport on the north-south axis, but it also became a barrier separating the city from the waterfront. While the list of stakeholders and institutions that were examined as part of this research (which includes the Washington State Department of Transportation, State Transportation Authority; City of Seattle and Seattle Dept. of Transportation; King County; Allied Arts (an NGO); Peoples' Waterfront Coalition (an NGO); Puget Sound Partnership (a third NGO)) does not claim to be exhaustive, it does include what were arguably the most prominent and vociferous voices involved in the debate.

In addition to identifying key stakeholders and institutions, attention must also be paid to the chronology of events that structure the narrative. Chronology is an important point to reflect upon as it establishes the flows of events and makes many things in the story comprehensible. However, the selection of events is inevitably subjective and reflects the value judgments of the researcher. Power issues and agendas may influence the narrator's account and selection criteria. This means that we may either face arbitrary and innocent-looking genealogies or carefully orchestrated stories. But even more interesting is the fact that one will have to recognize that a narrative chronology of events will have to include multiple phenomena and objects, some natural (e.g., an earthquake) and others of human origin (e.g., government bodies or stakeholder groups). The point is to include different realms such as the geology and tectonic plate movement, political-administrative systems, human stakeholders, the environment, and physical-material artifacts. Out of such complex assemblage grows the Seawall project. So, besides being a practical argument for listing a number of ontologically different objects within the same narrative chronology it is also indicative of the complexity of assemblages of artifacts and people that constitute a contemporary piece of urban infrastructure.

66   JENSEN

Date	Event/Element
1869	First recorded earthquake in the area.
April 4, 1953	Seawall section of State Route 99 opens.
1965	Major earthquake damages parts of the Seawall.
1973	City councilor John Miller identifies "the viaduct as the city's worst mistake."
1995	Engineers predict the viaduct will be unable to withstand a 7.5 scale earthquake.
February 28, 2001	6.8 scale Nisqually earthquake damages the viaduct.
2002	Conceptual engineering for a replacement begins.
2003	Peoples' Waterfront Coalition (NGO) is formed.
2004	The initial proposals are reduced to five options.
2005	A tunnel becomes the preferred option.
2007	An advisory vote is held in Seattle, both the surface-tunnel hybrid and an elevated structure receive a majority "no" vote.
2009	Governor, King County Executive, and Seattle Mayor recommend replacing the viaduct's central waterfront section with a bored tunnel (see http://www.wsdot.wa.gov/projects/Viaduct).
2010	Construction to replace viaduct's southern end commences.
2011	Beginning of the construction of the viaduct's central waterfront replacement (see http://www.wsdot.wa.gov/projects/Viaduct).
2013	New SR99 segment south of downtown opens to drivers.
2015	New SR99 segment through downtown Seattle opens to drivers.
2016	Removal of viaduct along central waterfront.

Figure 2: Timeline of selected events and human/non-human elements in the story of Seawall

In this chronology the idea has been to assemble human and non-human, physical, and institutional elements that affect the Seawall. The apparently arbitrary dimension to the list illustrates the importance of the storytellers and the selection criteria that they employ in its creation. In addition to the chronology of events presented here, it is worth stating that the Washington State Department of Transportation WSDOT developed a range of communication techniques which shaped the public dimension of the whole project. These techniques included site visits, blogging, and the creation of online "FAQs" (Frequently Asked Questions).

Citizens' Blogging and E-Mailing: The Narrative Reassembling of the Seawall

The WSDOT employed a number of media and fora to facilitate public debate about the project. Video streaming of hearings and public meetings, together with citizen blogs and guided tours of the site represent valuable sources of information. The blogs, in particular, allowed for alternative and/or critical narratives to be articulated:

> Lovely. Now when a breakdown happens it goes from 2.5 lanes to 1 lane. Did the politicians learn anything from I-5 downtown under the Convention Center? Reducing the lanes through downtown is the opposite of what you want to do! Plus the view is gone for us little people. How much is this going to cost us? $12 billion?! And just who will it benefit? Not the average Seattlite! Vote for a new mayor and new County Executive! This is ridiculous! At least with the replacement viaduct there were three lanes and a view. (Anonymous, WSDOT Blog, August 12, 2009)

> This entire thing is completely unAmerican. We voted AGAINST this solution and they are building it anyway. We are being lied to, robbed and utterly ignored. The land that will be opened up is being proposed to the public as public waterfront property, but in time it will be sold to private investors. It's time for a change in Washington State and Seattle. We need new politicians who will properly represent the public and what we the public vote on. (Anonymous, WSDOT Blog, August 12, 2009)

There are many more voices articulating criticism, support, or technical commentary on the blog site. The voices found on the Blog are not only those of "critical citizens," but also "professional testimonies" left by engineers and builders who tell "positive stories" about their engagement with the reconstruction project. Here I can only scratch the surface of a layer of the public sphere that is at work in this complex case. The sheer volume and complexity of documentation, information, and voices, however, makes an overview next to impossible. This, however, offers an important lesson about how infrastructure projects are assembled by multiple layers of physical as well as communicative networks.

Many other topics abound, but again this is not so much a case of showing the precise content of the Seawall process as it is an attempt to show the plethora of narratives unfolding from the event of the 2001 earthquake. Also, there are illustrations that emphasize how the Seawall assembles multiple publics, institutions, and technologies into one large, complex, socio-technical conglomerate with its point of departure in one physical place.

Inspection Processes: Citizen Tours as a Technique for Reassembling the Seawall

The physical site of the Seawall project was visited on a regular basis by a number of citizen tours organized by the WSDOT. Advertisements invited interested parties to:

> Join us for a tour of the Alaskan Way Viaduct on Saturday, March 27 [2010].... Tours will be held between 9:30 a.m. and noon. If you would like to participate, e-mail viaduct@wsdot.wa.gov or call 1-888-AWV-LINE and leave a message with your contact information. (http://www.wsdot.wa.gov/projects/Viaduct)

Another technique which attempted to facilitate public debate was the FAQ section of the WSDOT website, which provided information about which parts of the road would be replaced, the viability of replacing the Seawall with alternative structures, and the likely costs associated with the project. Through this portal, the Seawall became embedded into a complex web of intertextual linkages which offered citizens the opportunity to access additional information. One question made reference to Boston's infamous "Big Dig," in which a section of Interstate 93 was diverted underground. The "Big Dig" was one of the most expensive and lengthy highway projects in recent U.S. history and is often cited as an example of deterrence in urban infrastructure matters.

In addition to public views, a long series of official and unofficial documents were produced in support of and in opposition to the proposed redevelopment scheme. In the document, "Waterfront for All," published by the NGO, Allied Arts, in 2006, the Seawall project is addressed with history as the point of reference:

> Fifty years ago, our civic leaders made a serious mistake. They cut off Seattle from its waterfront by building the Alaskan Way Viaduct. Now, the people of Seattle and the Northwest have an opportunity to correct this error and redirect the future of the region. We have the choice of giving future generations a vibrant Waterfront neighborhood, or cursing them with an even larger viaduct ripping through some of the most significant urban land in the Northwest.... Allied Arts stands with countless other civic and community organizations that believe our new Waterfront is the best opportunity we'll have to maintain and enhance our region's quality of life for the foreseeable future." (Allied Arts, 2006, p. 4)

The document is special in the sense that it also contains a number of colorful images that illustrate the architects' and urban designers' imagined futures for the waterfront. This "rhetoric of illustrations" is interesting to add to the many verbal contributions about what should be done as well as illustrating a different dimension of the power of representation than the many rational and quantitative data related to engineering, economics, and traffic flows.

The official City Government document, "Waterfront Concept Plan," carries the authoritative note on the front page stating that this is "the Mayor's recommendation." The document was issued in 2006 and in it there is an explicit perception of the meaning of mobility to the imagined future scenario under the heading of "Movement":

> Seattle's waterfront is a place of movement. Pedestrians, bicycles, cars, trucks, streetcars, trains, ferries, water taxis, cruise ships, and more are continuously and simultaneously moving about the waterfront 24 hours a day, seven days a week. Pedestrian movement on the Waterfront is a fundamental activity for relaxation, health, and enjoyment of the waterfront's, public spaces, art, views, landmarks and shoreline. As people move between the city uplands and the waterfront, their experience is one of viewing landmarks in sequential relationship to each other. (City of Seattle, 2006, p. 15)

The Seattle Department of Transportation (SDOT) published its comprehensive *Urban Mobility Plan* in January 2008. The introduction stated that:

> The Urban Mobility Plan (UMP) is an opportunity to ensure Seattle's Center City will continue to grow in size, economic vitality and accessibility because existing infrastructure is made more efficient, inviting, and accommodating. The Plan also recognizes the importance of the effective movement of goods, protection and support of industry, facilitation of Port activities, and continued attraction of large and small business. The Nisqually earthquake reinforced the need to look at alternatives to the current Alaskan Way Viaduct, which divides Seattle's waterfront from its downtown core." (Seattle Department of Transportation (SDOT), 2008, p. 1A-3)

The Seawall thus became a vehicle for articulating different hopes and visions for the future of Seattle, as well as for the more mundane ambition of shutting up one's political opponents. One particularly important voice belonged to Mayor Greg Nickels. Mayor Nickels appeared particularly keen on the "making history" and "new beginning" discourses that a number of people gave voice to (Mayor Greg Nickels, Mayor's Press Conference, December 6, 2004). In this political rhetoric, the city is not only facing new beginnings that may offer the window of opportunity to create a new infrastructure, it also represents a point in time where the inhabitants of Seattle can become united by new challenges:

> The City of Seattle approaches this issue united. Even though we are obviously a democratic community and we will have lots of voices speaking to this issue in different perspectives in different ways, the Seattle City Council has been a partner on this

the whole way . . . it's important that we speak with one voice on an issue of this magnitude with our transportation future. (Mayor Greg Nickels, Mayor's Press Conference, December 6, 2004)

The voices of power do not stand uncontested though. An example is found in this article from December 9, 2004 in which Cary Moon, the editor of the local weekly, *The Stranger*, and a vocal Seawall critic, wrote against the notion of Nickels's big unifying event. Moon wrote this under the headline title "A Bad Case of Tunnel Vision":

> I think Mayor Nickels has underestimated how deeply Seattleites care about creating a sustainable urban future, and how many of us probably actually would prefer a simple and affordable solution to an expensive, complex megaproject. So why do project officials continue to ignore this potential solution that costs a fraction as much as the tunnel, avoids the costs and risks associated with megaprojects (Hello! Boston?), could make the whole system function better, and may offer a far superior economic payoff? Clearly, there is a poetic and powerful vibe emanating from the gray, hulking mass. Anyone tapped into the layered history of our city knows that the grit and decay of this place is a rare bit of this essence and soul left downtown. But the beast is dying; there's no reviving it. Let's say our goodbyes, have a wake, and close that chapter of our history forever. (Moon, 2004, n.p.)

The Mayor also had to address more difficult issues in 2005 when he had to deal with the fact that there was "no pot of local money" large enough to finance the rebuilding of the Seawall (Mayor Greg Nickels, Mayor's Press Conference, April 26, 2005). Clearly such appeals to higher tiers of government and funding are seen in most debates over funding costly infrastructure. Seen from the vantage point of the relational and mobility-defined sense of place presented earlier in this chapter, this makes sense, as a city like Seattle is inexorably bound into regional and national networks of mobility that local contributions alone could not afford to build or maintain. The key issue here is that the Seawall is a State Road. A political-administrative assemblage like the Seawall immediately draws upon multiple networked relationships that cross local administrative boundaries and transgress government tiers.

In addition to these powerful stakeholders were a number of NGOs. One of the most influential was the Allied Arts organization, whose mission is to "enhance the cultural livability of Seattle and to create a social network of people who care about the Arts, Urban Design and Historic Preservation" (http://www.alliedarts-seattle.org/). This group also refers to the "making of history" and the earthquake as something that transformed a destructive event into a positive opportunity for change that made a "new beginning" possible. Likewise the People's Waterfront Coalition (PWC) is another important NGO that attempted to influence the debate about the waterfront. The PWC advocated traffic solutions that would promote increased use of public transit systems, walking, and cycling. They, too, used the earthquake to articulate discourses about "new beginnings." Julie Parrett, a PWC member, stated that for 50 years the transportation solution in cities across the US had been to build highways, and provocatively argued that "Seattle can live without the Viaduct" (http://www.peopleswaterfront.org/who.html).

Listening to the Seawall

Here I will intentionally bypass standard academic conventions to offer an imagined dialogue as it might have sounded "if only the Seawall could speak." Clearly this approach violates established criteria for "objective research." However, the academic references and the empirical data and references used in this story so far do bear witness to an attempt to present the case from a vantage point which is arguably closer to people in Seattle than my own. The attempt to avoid idiosyncratic interpretations and pet issues has been central. However it is a key feature of the "narrative" approaches to social research that one needs to rid the illusion of "pure objectivity." Thus Czarniawska (2004, p. 121) borrowed Bakhtin's concept of "heteroglossia" [many voices] as an attempt to acknowledge the plethora and multiplicity of voices. Accordingly, a polyphonic ethnography that recognizes multiple voices resists the illusion of having people speaking for themselves. There is always an element of "authorial strategy" orchestrating the many voices of a narrative like the one of the Seawall.

Furthermore, Czarniawska argued that social scientists do most harm when they impose their interpretations of what they argue are "authentic voices of the field" (2004, p. 122). I cannot enter the epistemology debate on where precisely to draw the boundary between the researcher and the voices in the field. However I have applied the conventional tools of research, such as accurate quotation and detailed referencing, in presenting statements that are not mine. Inspired by Latour's (1996) attempt to blur the distinctions between the social and the technological, I have been searching for a way of condensing the argument and presenting the topic in a more provocative manner. Latour spoke of inventing a hybrid genre he called "scientification" in which he claimed to fuse science fiction, fiction, realism, journalism, and human sciences (Latour, 1996, pp. viii–ix). While I cannot claim to have Latour's originality, I attempt to adopt a similar approach when considering the Seawall. The status of the following imaginary dialogue with the Seawall must therefore not be confused with a claim to objective research and representation. Neither is this an advocacy for sheer imaginary inventions as representational format in social research. Rather it is an attempt to articulate the view of complexity of contemporary infrastructures as being assembled by much more than just asphalt and steel.

If we imagine that the Seawall can "speak" we might be able to better articulate the multiplicity and complexity of the development as seen from an illusionary (and fictitious) vantage point. Furthermore, experimenting with this format exposes the problem of research reporting as something neutral in which researchers apply an "objective style" of writing which elevates the human observer while ignoring the presence of objects in the text (Latour, 2005). It also becomes increasingly difficult to insist on subjects and objects being hermetically sealed off from one another. Rather, people act by means of objects of all sorts and most of what we study within empirical social research does not make sense if the materiality of the multiple interactions and networks are not included. The imaginary dialogue and the ascribing of a voice to an inanimate object is thus an attempt to challenge this separation of subject and object, of nature and culture, of the social and the technical. Even though the Seawall cannot speak, imagining that it can and giving it voice moves beyond the humorous play of a single sociologist. It touches upon one of the greatest challenges of contemporary social research: how to understand the complexity of networks and agents in multiple assemblages of technology and sociality. I cannot claim to have solved that problem but I consider this exercise a way of raising the issue of how the "sociology of associations" may assemble that which has been suspended by the imagined coherence and homogeneity of the "social" and "society."

The following imaginary dialogue is thus both an attempt to draw on the work of Latour (1996, 2005), and also an effort to take a point of departure in the contemporary critique of uncritically inheriting notions such as "society" (Urry, 2000) and "representation" (Thrift, 2008). Finally, I consider the little exercise of imagining a dialogue as a parallel to the well-established procedure of the "thought experiment" and the posing of the question, "what if...?" Having outlined the rationale for this rather unorthodox approach I find it time now to turn to the Seawall. So let us for a brief moment imagine what the Seawall would say, if only it could speak. I have chosen to represent this as an imaginary dialogue between a researcher and the Seawall. Thus, imagine the researcher talking to the Seawall.

> Researcher: So tell me, what actually happened that day back in 2001?

> Seawall: Seriously, how much time have you got? Some think it all started with the shaking of the earth on that Wednesday back in February 2001, but actually they had started inspecting me and discussing my physical condition long before. I suspect they had plans for my future in the pipeline then. But you're right, the earthquake changed

things. Not least in the public debates. Suddenly I was the window into a sea of change, a new tomorrow and all that. . . .

Researcher: How did it feel to become the center of so much attention in the city?

Seawall: It is kind of odd. I had been here for some time, but after I got real sick it has been amazing to hear how many people are drawing me in multiple different directions and how they evoke emotions and feelings that are hurtful. How would you feel if you had carried thousands of vehicles throughout decades and then when you get ill you must listen to people calling you all sorts of negative things and some even arguing that you were a disgrace to the city? Moreover, it is really not nice to see how some people actually lie and argue from points that cannot claim to have any relation to what actually happened. It seems like humans are very busy putting each other down but at the same time they sound rather oblivious to that fact that they are fully dependent on technologies and objects.

Researcher: Hmm, I sense a certain disappointment in your voice. Are you thinking different about us humans after your recent experiences?

Seawall: There's been a lot of talking. There are a lot of words. I don't mind. I guess that is only natural. I just don't understand how humans can fail to see that there are important elements of an urban intervention that is not derived from their immediate actions and doings. We are actually a very large number of objects and things in the world that are the preconditions of contemporary urban life, regardless if the self-propelling human agents intervene or not. Life without objects would mean a very different life both socially and aesthetically as well as in political-economic terms you see. Humankind has experienced this before, but it seems such a long time ago that they have forgotten what pre-technological life felt like . . . if a thing like that ever existed. I guess what I am trying to say is . . . just don't forget how we shape and form your life!

Concluding Remarks

Of course the Seawall cannot speak. But from this exercise it seems that I may enhance our understanding of how complex the relationships between human agents and material artifacts are. If I had written a narrative about how a car, or a personal computer, would have felt, the reader might have understood and accepted it more easily, since it seems that within the technoculture we inhabit humans more readily bond emotionally with some machines and artifacts than with others. But of course they cannot speak either. Here the key point is to open up our understanding of road infrastructures as physical entities that go way beyond just fulfilling their utilitarian goals of facilitating mobility from point A to point B (Jensen, 2009). The Seawall (and indeed any other large urban structure) is an evocative and cultural artifact that becomes taken for granted, but which also becomes a site of contestation as soon as an event opens up debates concerning what it actually means to the city and its inhabitants.

Obviously, the thought experiment makes no claims to scientific or objective rationality, but it does offer another angle on the complex issue of human/non-human interaction. By asking the "what if?" question we become alert to the "thinking" and "speaking," non-sentient objects. This allows us to recognize that alternative perspectives may be necessary for us to understand the importance and meaning of non-human agency and control. As the Seawall reminded us, "life without objects" and a "pre-technological life" is not an option even though our understanding often freezes into an anthropocentric mindset. Many lessons can be taken from the story of Seattle's Seawall. One is the importance of understanding how an earthquake is more than a natural "event" (for a discussion of mobility, natural events, and socio-technical systems, see Jensen, 2011), as its meanings and repercussions are deeply embedded into the socio-material fabric of the city. Furthermore, in exploring the narrative of the Seawall one finds a story about utility and the instrumental organization of flows, but also about culture and inhabiting place via an understanding of the meaning of movement. The concepts of relational linkages and mobility enable us to see the Seawall as a material and cultural artifact that assembles multiple voices from the field, thus constituting a specific notion of place. Theoretically and methodologically this raises interesting challenges to urban studies as well as to technology and cultural studies.

As always, large infrastructure investments and interventions pitch environmentalists versus economic growth advocates, car-oriented commuters versus walking and cycling enthusiasts, and aesthetic considerations against efficient transportation logistics. This is also the case with the Seawall in Seattle, and the development controversy would have needed much more careful and systematic investigation if the aim of the chapter had been to assess what the best or "correct" decision was. My concern, however, has been emphasizing the ways in which stakeholders and institutions blend with hardware and technology to create huge, socio-technical complexes. To return to the point from the introduction: one may understand infrastructure as a relational assemblage of "hardware" (e.g., asphalt, concrete, columns, traffic lights, etc.) and "software" (e.g., strategies, traffic codes, safety campaigns, etc.). It is this nexus of material and immaterial, human and non-human that needs to be understood in relation to the flow and friction it affords. While the imaginary dialogue between the Seawall and a researcher can be considered something of a gimmick, the thought experiment does lend itself to some of the more fundamental questions that arise from Science and Technology Studies and Actor Network Theory (see, for example, Farias & Bender, 2010; Hommels, 2005; Latour, 2005).

This chapter cannot claim to contribute to either of these consolidated fields, but what it can do is to open up perspectives and represent urban interventions in a narrative and discursive format with concerns for the notion of mobility and place. "If only it could speak" would enable us to listen to exciting stories about power and politics, but would also further reinforce just how dependent we are on technologies and hardware that are seemingly under societal control. There is potential here for combining narrative theories and the mobility turn. It is difficult to predict where this will lead us, but such an approach offers an invitation to explore notions of narrative, mobility, and place as seen through the story of Seattle's contested Seawall.

Notes

1. A form of speech in which an imaginary, dead, or absent person speaks.
2. A term devised by Latour to refer to a hybrid genre which fuses science fiction, realism, journalism, and human sciences.

References

Allied Arts. (2006). Waterfront for all. Retrieved from http://www.alliedarts-seattle.org/, http://www.alliedarts-seattle.org/index.php?option=com_content&task=view&id=11&Itemid=27, and http://www.alliedarts-seattle.org/images/download/AlliedArtsFull.pdf

Anderson, R. (2002). The viaduct at a crossroads: Dutiful servant, brutal barrier. *The Seattle Times*. Retrieved from http://seattletimes.nwsource.com/pacificnw/2002/0407/cover.html

Beauregard, R. A. (2005). From place to site: Negotiating narrative complexity. In C. J. Burns & A. Kahn (Eds.), *Site matters: Design concepts, histories, and strategies* (pp. 39-58). London, UK: Routledge.

City of Seattle. (2006). Waterfront concept plan. Retrieved from http://www.seattle.gov/dpd/cms/groups/pan/@pan/@plan/@proj/documents/web_informational/dpds_008310.pdf and http://www.seattle.gov/dpd/planning/central_waterfront/archive/draftwaterfrontconceptplan/default.asp

Cresswell, T. (2006). *On the move: Mobility in the modern Western world*. London, UK: Routledge.

Czarniawska, B. (2004). *Narratives in social science research*. London, UK: Sage.

DeLanda, M. (2006). *A new philosophy of society: Assemblage theory and social complexity*. New York, NY: Continuum.

Dorpat, P., & Crowley, W. (2000). Seattle central waterfront tour. Retrieved from http://www.historylink.org/index.cfm?DisplayPage=output.cfm&file_id=7072

Dovey, K. (2010). *Becoming places: Urbanism/architecture/identity/power*. London, UK: Routledge.

Eckstein, B., & Throgmorton, J. A. (Eds.). (2003). *Story and sustainability: Planning, practice, and possibility for American cities*. Cambridge, MA: MIT Press.

Farias, I. (2010). Introduction: Decentering the object of urban studies. In I. Farias & T. Bender (Eds.), *Urban assemblages: How actor-network theory changes urban studies* (pp. 1-24). London, UK: Routledge.

Farias, I., & Bender, T. (Eds.). (2010). *Urban assemblages: How actor-network theory changes urban studies.* London, UK: Routledge.

Finnegan, R. (1998). *Tales of the city: A study of narrative and urban life.* Cambridge, UK: Cambridge University Press.

Flyvbjerg, B. (1998). *Rationality and power: Democracy in practice.* Chicago, IL: University of Chicago Press.

Graham, S., & Marvin, S. (2001). *Splintering urbanism: Networked infrastructures, technological mobilities and the urban condition.* London, UK: Routledge.

Hommels, A. (2005). *Unbuilding cities: Obduracy in urban socio-technical change.* Cambridge, MA: MIT Press.

Ingersoll, R. (2006). *Sprawltown: Looking for the city on its edges.* New York, NY: Princeton Architectural Press.

Jensen, O. B. (2007). Culture stories: Understanding cultural urban branding. *Planning Theory, 6*(3), 211-236.

Jensen, O. B. (2008, November). *European metroscapes: The production of lived mobilities within the socio-technical metro systems in Copenhagen, London and Paris.* Paper presented at the *Mobility, the City and STS* Conference, The Technical University of Denmark (DTU), Copenhagen, Denmark.

Jensen, O. B. (2009). Flows of meaning, cultures of movements—Urban mobility as meaningful everyday life practice. *Mobilities, 4*(1), 139-158.

Jensen, O. B. (2011). Emotional eruptions, volcanic activity and global mobilities—A field account from a European in the US during the eruption of Eyjafjallajökull. *Mobilities, 6*(1), 67-75.

Kolb, D. (2008). *Sprawling places.* Athens, GA: University of Georgia Press.

Latour, B. (1996). *Aramis, or, the love of technology.* Cambridge, MA: Harvard University Press.

Latour, B. (2005). *Reassembling the social: An introduction to actor-network-theory.* Oxford, UK: Oxford University Press.

Lefebvre, H. (1991). *The production of space* (D. Nicholson-Smith, Trans.). Oxford, UK: Blackwell.

Moon, C. (2004, December 9). A bad case of tunnel vision. *The Stranger.* Retrieved from http://www.thestranger.com/seattle/Content?oid=20044

People for Puget Sound. (2009), Urban waterfronts in Puget Sound. Retrieved from http://pugetsound.org/policy/issues/urban-waterfronts, and http://pugetsound.org/, or http://www.pugetsound.org/

People's Waterfront Coalition. (2006). Citizens for a highway-free shore. Retrieved from http://www.peopleswaterfront.org/

Puget Sound Partnership. Puget Sound vital signs. Retrieved from http://www.psp.wa.gov/

Richardson, T., & Jensen, O. B. (2008). How mobility systems produce inequality: Making mobile subject types on the Bangkok Sky Train. *Built Environment. 34*(2), 218-231.

Scollon, R., & Scollon, S. (2003). *Discourses in place: Language in the material world.* London, UK: Routledge.

Seattle City Council. (2011). Alaskan Way viaduct and Seawall replacement project and central waterfront planning. Retrieved from http://www.seattle.gov/council/issues/viaduct.htm, and http://www.seattle.gov/council/, or http://www.seattle.gov/council/video_archives.asp

Seattle Department of Planning and Development. *Waterfront Seattle*. Retrieved from http://www.seattle.gov/DPD/Planning/Central_Waterfront/Overview/

Seattle Department of Transportation (SDOT). (2008). *Urban mobility plan*. Retrieved from http://www.seattle.gov/transportation/docs/ump/01%20SEATTLE%20introduction%20to%20plan.pdf

Thrift, N. (2008). *Non-representational theory: Space, politics, affect*. London, UK: Routledge.

Turkle, S. (1984). *The second self: Computers and the human spirit*. New York, NY: Simon & Schuster.

Urry, J. (2000). *Sociology beyond societies: Mobilities for the twenty-first century*. London, UK: Routledge.

Vannini, P. (2008). A queen's drowning: Material culture, drama and the performance of a technological accident. *Symbolic Interaction, 31*(2), 155-182.

Washington State Department of Transportation. *SR 99–Alaskan Way Viaduct replacement*. Retrieved from http://www.wsdot.wa.gov/projects/Viaduct/

5

How Car Drivers Took the Streets: Critical Planning Moments of Automobility

Nick Scott

Canada has about 900,000 kilometers of road, enough to circle the globe 22 times (Transport Canada, 2010). Lying sturdily beneath us, roads seem like they were always there, and barring bad weather or subterranean pipe failures, we take them for granted. Once we think about their variety, however, some important differences arise. For example, not all roads are for all people, despite a historical role of roads as public space (Van Nostrand, 1983). Which people and machines can use urban roads cuts to the heart of how and for whom cities are produced and how they impact upon the environment. In this sense, roads are not passive, geometric backgrounds against which the dynamic events of urban history unfold. Rather, roads constitute a technology of mobility (Hommels, 2008; Jensen, this volume; Norton, 2008). As such, they become attached to particular values, practices, institutions, and resource chains in the production of space. If a particular kind of road becomes obdurate, or difficult to change, this has as much to do with people and ongoing power relations as it does with bituminous glue and concrete cement.

In what follows, I explore the production of space for car driving in the North American context. I argue that automobility significantly affects the production of space and that roads figure centrally in this process. Automobility refers to the networks of natural resources, cultural values, political relations, and techno-sciences which, in concert with one another, afford the widespread practice of automobile driving (Dennis & Urry, 2009). The production of space, following Henri Lefebvre (1991), refers to not merely a technical exercise, but an interplay between urban planning, ideas about the "good city" (Amin, 2006), and spatial practice. I begin by considering how car driving can be understood as a civic practice that affects the nature of urban community. In the second section, I examine how automobility affects the production of space through the planning process and the organization of material infrastructure. In the third section, I mobilize my central argument that the car's influence on the city can be illustrated through three critical "planning moments" of automobility. These moments are characterized by novel associations between the car and urban space, and conflicting visions of the good city.

Car Driving as Civic Practice

The automobile weighs heavily on North American society, profoundly affecting the way people live together. A diverse set of actors, from governments and planners to consumers, have turned the city into a car-dependent place, tying freedom of movement to the systemic practice of car driving (Jones, 2008; Urry, 2004). Recent research on automobility explored how car networks have become "locked in." This research balanced the private benefits of car driving against the collective threats that mass car travel poses to global ecology, public health, and social justice (Conley & McLaren, 2009; Paterson, 2007). A lesser explored area is whether automobility promotes liberal notions of freedom, such as the autonomy to choose where and how one wants to move through space; an association mass advertising has made since the car's inception. For example, Rajan questioned whether the "freedom" offered by automobility is actually a "compulsory constraint" (2006, p. 123). Another area of political thought that could be further developed in relation to automobility is citizenship theory (Kymlicka, 2002), including the argument that certain civic practices are vital for the "good city," or "the kind of urban order that might enhance the human experience" (Amin, 2006, p. 1009). Exploring everyday car driving as a civic practice that shapes the nature of urban community presents one way of situating the car within the social and political production of space.

Interest in citizenship practices has grown in recent years in the context of long-term voter apathy, growing inequality, and political hostility towards welfare and multiculturalist policies (Kymlicka & Norman, 2000). Citizenship theory, according to Kymlicka (2002), describes an attempt "to integrate the demands for liberal justice with community membership" (p. 284). It argues that in addition to civil, political, and social rights tied to the state (Marshall, 1950), stable liberal democracies require active public engagement, including the everyday practice of civility and solidarity (Galston, 1991). A sense of civility, for instance, relates to "the way we treat non-intimates with whom we come into face-to-face contact" (Kymlicka, 2002, p. 301), a form of contact which has become increasingly mobile (Amin & Thrift, 2002, p. 38). Like civility, solidarity becomes relevant during many routine activities that define ordinary life, such as buying groceries, driving a car to work, and managing social relationships. Not all forms of solidarity are inclusive or democratic. A city in which diverse groups of people can flourish requires solidarity with the stranger and the outsider, or what Dean (1996) called a "reflective solidarity" based on an "awareness of and regard for those multiple interconnections in which differences emerge" (p. 16). Similarly, for Amin (2006) a habit of solidarity among strangers can help bring about the good city by prodding public culture "towards outcomes that benefit the more rather than the few, without compromising the right to difference that contemporary urban life demands" (p. 1012).

In this context, I argue, car driving constitutes a civic practice that creates an automobile citizenry, and figures centrally into how and with whom people participate in urban social life, starting at an early age (Packer, 2008). As Sheller (2004) argued, the car has become "deeply integrated into the affective networks of familial life and domestic spaces" (p. 230), including the journey to school and after-school activities. Frequently borne by mothers, driving children to school is construed with safety and "good mothering" (Murray, 2008, p. 53). Constant chauffeuring enables group participation, but also deprives children of outdoor time to actively and independently explore their surroundings, a systemic problem often overlooked by local educational programs (Parusel & McLaren, 2010). In Canada, when teenagers turn sixteen they can apply for their own driver's license, two or three years before they can purchase alcohol and participate in electoral democracy. As an obligatory rite of passage wrapped up with other milestones such as sexual activity, a driver's license constitutes a significant amount of power for young adults. Best (2006) showed that "car privileges," for example, not only represent freedom from prescribed roles at home and school, but also offer a way to economically support family members, negotiate evolving identities, and participate in wider civil society and public life. Becoming a car driver, therefore, involves the development of skills that go well beyond operating a motor vehicle. It forms part of a civic education in which people learn "how they fit together with others" (Taylor, 2004, p. 23) and how to participate within "civil societies of automobility" (Sheller & Urry, 2000, p. 739).

Automobility enables civic participation but also produces socio-spatial exclusion (Preston &,Raje 2007), suggesting an ambivalent relationship with the good city. In North America cities are structured through asymmetrical power relations that privilege the participation of motorists, an imbalance which remains remarkably stable as it cuts across the myriad other identities, cultures, and communities with which people affiliate. This stability creates a source of tension and conflict in the city. Although the majority of Canadians drive in order to get to work and reach most of their daily destinations (Turcotte, 2008, p. 22), automobility is neither universally accessible nor an equally affordable mode of "compulsory consumption" (Soron, 2009). Furthermore, mass automobility depends on the interactive inequality, spatially embedded, between car drivers and others—when and where others are without cars. Cyclists and pedestrians, in particular, lacking air cushions and steely exoskeletons to dampen a collision, raise the stakes of civility and solidarity on urban roads (confirmed by the casualties of traffic faithfully chronicled on public radio) (Furness, 2010; Short & Pinet-Peralta, 2010; see also Conley, this volume).

Motorists periodically become "monsters in metal cocoons" (Lupton, 1999). However, much of the incivility associated with car driving occurs indirectly, owing to a systemic disconnect between the personal experience and benefits of car driving and the toll it takes on others through greenhouse gas emissions, noise, ecological fragmentation, and the noxious particulates that exit tailpipes (Brugge, Durant, & Rioux, 2007). As Vanderbilt (2009) lamented, "we do not pay for the unsavory emissions our cars create" (p. 160). Overall, the power relations which privilege the practice of car driving suggest that automobility plays a significant role in assembling cities. In the following section, I consider the planning process by which automobility influences urban space.

Cars and the Production of Space

Automobility may shape urban community, even more so than through any one particular practice or institution, through the production of space. According to Lefebvre (1991), the production of space involves an attempt to implement abstract representations or "conceptualized space, the space of scientists, planners, urbanists, technocratic subdividers and social engineers, as of a certain type of artist with a scientific bent" (p. 38). When conceptualized space is applied to the built environment, it interacts unevenly with established practice, reordering "the routes and networks which link up the spaces set aside for work, 'private' life, and leisure" (1991, p. 38). The technical and objective language that planners and engineers often use is therefore belied by the inherently political character of the planning process (Flyvbjerg, 1998). Some conceptualized spaces, as Lefebvre decried, have been used to dominate and even destroy complex urban fabric for sake of an abstract order fixated on narrow visions of growth (Scott, 1998). For example, during an era of superhighway fever in the 1960s and 1970s freeways were punched through established neighborhoods, usually poor and housing minority racial groups (Bullard, Johnson, & Torres, 2004). Freeways, however, were not a product of early 20th century planning. Rather, they reflect the culmination of a dramatic shift, starting in the 1920s, towards functionalist representations of urban space that were preoccupied with accommodating the uncontested circulation of private automobiles (Brown, 2005). This shift, I will argue below, marks a critical "planning moment" of automobility. In spite of political resistance (Jacobs, 1961), the reconstruction and expansion of freeways has become a salient component of transportation planning (Brown, 2006).

Planning involves drawing boundaries around communities, drawing from the larger cultural and commercial context of which it forms a part (Jensen, 2007). As planners, engineers, publics, and a diverse set of profit-seeking actors attempt to create a particular vision of space, they engage in a spatial politics of inclusion and exclusion in which some spatial practices are privileged over others (Routledge, 2010). Mobilities planning, for example, entails a relational process of "pacemaking" wherein speed, slowness, and risk are unevenly distributed (Hubbard & Lilley, 2004). In the 20th century this distribution came to increasingly favor motorists as planning privileged and found legitimization in the car (Gordon, 2001; Norton, 2008). Car culture, as Dennis and Urry (2009) suggested, "has developed into a dominant culture generating new ideals about what represents the 'good life' and what is necessary to be a good mobile citizen" (p. 37). As a result, the ideals of car culture slip seamlessly into planning narratives and overshadow the ideals in other cultures of mobility. According to Eckstein (2003), plans can be judged by how effectively they create "space amenable to multiple stories, how well the arrangement of that space produces provocative interaction among the stories, and thus how well and how broadly the stories are heard" (p. 22). Similarly, lived spaces might be judged by the extent that they facilitate civil encounters with difference, where people move outside of "familiar enclaves" and meet strangers (Young, 1990, p. 397). As the car driver developed into a focal actor of planning narratives, the spatial perspectives of pedestrians, cyclists, and transit users became excluded from city growth (Conley & McLaren, 2009; Norton, 2008). Increasingly, the roadways and powerful metal prosthetics to which drivers delegate the work of automobility come between urban encounters with difference (Thrift, 2004).

Material infrastructures assembled through the planning process also play a significant role in the integration of car drivers into the city. Like the hybrid car driver, the "auto-space" in which the car driver is grounded (Beckmann, 2001, p. 603) can be viewed as an urban assemblage in which human and nonhuman actors align through a process of "translation" (Farías & Bender, 2010). In actor-network theory, translation means the creation of linkages between heterogeneous entities and emphasizes the agency of material objects as "mediators" in sociotechnical networks (Latour, 2007, p. 108; Law, 1999). For example, mediators include the sewers, hydrocarbons, lights, medians, sidewalks, and snow plows that help facilitate seamless car trips. When auto-space works smoothly, the motorist "will find no difficulty in distinguishing what is displaced from the immutable framework in which it is displaced" (Latour, 1997, p. 174).

Seamless auto-space engenders two problems, however, for automobile civil societies. First, it compounds the separation car drivers already experience from their surroundings while inside "sonic envelopes" (Bull, 2004). As Urry (2006) argued, while "dwelling at speed, car drivers lose the ability to perceive local detail, to talk to strangers, to learn of local ways of life, to stop and sense each different place" (p. 23). Second, a seamless passage for motorists depends on a sprawling infrastructure that severs the spatial linkages that make alternative mobilities possible, entrenching the marginalization of walking and cycling (Sheller & Urry, 2000). As this infrastructure is locked into place by sunk costs, fixed assets, and a vast "maintenance constituency" (Staudenmaier, 1985, pp. 195-196), it becomes difficult to reassemble. Closely interrelated with many other elements of the city, from residential construction and work routines to waste water systems, roads networks produce a high level of what Hommels (2008) called "relational obduracy" (p. 27). Adapting one element of a complex network requires changing many others.

The manner in which car driving affects the production of space can be illustrated, I argue, by three critical "planning moments" of automobility. Automobility and planning do not constantly engage in a dynamic way, when far-reaching, political contingencies become visible and vulnerable to change. Periodically, however, they may establish a novel relationship, or in some cases, undergo a fundamental relational shift. The core idea that I want to capture in these three critical planning moments relates to transformations in what Cresswell (2010) called "constellations of mobility." According to Cresswell, "historically and geographically specific formations of movements, narratives about mobility and mobile practices" create constellations that characterize particular spaces of mobility (2010, p. 17). Changes in such constellations could also be described as changes in actor-networks, where the introduction of associations with new actors transforms the network (Callon, 1989, p. 93). A second feature follows closely from this idea of transformation. Namely, during a critical planning moment, multiple visions of the good city converge and work together to reshape spatial practice and material infrastructure. These visions grow out of the salient planning movements of the 20th century and reinforce the notion that car driving constitutes a civic practice deeply implicated in the construction of community life (Hall, 2002). To be sure, a multitude of planning moments exist in which the car modified urban space, and not all of the ways in which the car influences the city relate directly to urban planning. However, I maintain that each case which I qualify as critical stands out because of the number and kind of associations established during the moment in question between previously unconnected actors. Equally important are the differences between the three planning moments which emphasize significant divergences in the way cities can facilitate automobility.

Critical Planning Moments of Automobility

The plurality of automobility becomes clear, if instead of a monolithic trajectory of mass car travel, we focus on a few critical junctures across specific jurisdictions in which automobility interacts dynamically with planning. To this end, I turn now to three critical planning moments seen through the lens of Canada's capital city and southern Ontario, the founding region of Canadian automobility. Ottawa presents a strategic site because all three moments were generated in the past and persist in some form today. As expressions of national power and identity, capital cities, according to planning historian, David Gordon (2002), "were often the location of early experiments in urban design, parks, public health, and social reform, as a broad movement to establish urban planning emerged in Europe and North America" (p. 30). I refer to the first critical moment as *park roads*. Shaped by planning movements which predate mass car travel, this moment marks an intriguing effort to insinuate early car drivers into their natural surroundings. The second moment, *machine roads*, entails a radically different conception of car space, closely aligned with spatial practices and materials of the sprawling post-war metropolis. The third critical moment is *new urban roads*. Like machine roads, the defining representations of new urban roads are currently supported by urban regimes across the continent which continue to negotiate their application. These moments do not correspond with sharply defined blocks of time. Rather, they constitute overlapping durations in which new representations of roads are translated into spatial networks that transform existing built environments. A moment comes to a close as newer representations emerge and through translation start to add further "sedimentary layers" (Lefebvre, 1991, p. 229). By reading across these three critical planning moments, I contend, in each of which the car and city relate in a novel way, we can grasp the far-reaching impact of automobility on the production of space.

The first moment: Park roads

In the early 20th century, three distinct planning movements, each a unique reaction to the ills of the Victorian slum city, offered competing visions of urban reform which shaped the introduction of automobility. The first was the Garden City movement. Inspired by the radical ideas of Ebenezer Howard to build self-sufficient cities in the countryside, surrounded by large greenbelts, the Garden City was as much a model for self-government and cooperative enterprise as it was a "back-to-land" movement (Hall, 2002, pp. 93-95). The second approach, City Beautiful, was more authoritarian. Its prophet, Daniel Burnham, famously implored in his 1909 blueprint for Chicago to "make no little plans. They have no magic to stir men's blood." Following Haussmann's reconstruction of Paris, this movement employed the assumption that a beautiful city, composed of radiating boulevards, parks, and monuments, would make its citizens better people (Hall, 2002, pp. 190-196). Finally, the City Scientific approach, which became the dominant mode of planning in Canada after WWI (Gordon, 2008), prioritized healthy circulation. Allied with civil and railway engineers, City Scientific planners were concerned with improving the social and technical efficiency of urban infrastructure, such as roads, water mains, and sewers, through new materials and construction techniques. Principles from each of these planning movements, Garden City, City Beautiful, and City Scientific, contributed to a novel representation of space: the park road.

In the first critical planning moment of automobility, space for the motor car was cut from the same cloth as the park. A park road is a "pleasure road," defined by the experience of wending through a natural landscape crafted to coincide with the motorist's visual perspective. Frederick Law Olmsted introduced the first parkway system to North America in Buffalo during the 1890s. He believed that "a park road is pleasant by reason of that which adjoins it, or is open to contemplation from it, not because it favors speed" (Olmsted, 1997, p. 257). A protégé of the Olmsted office, Frederick G. Todd, became Canada's first resident landscape architect, based in Montreal, and in 1903 brought with him to Ottawa a Garden City philosophy of putting road users in thrall to their natural surroundings. "The term Parkway I have taken to mean a winding pleasure drive laid out with a narrow strip of land reserved on either side, and treated in a park-like manner," an exemplar of which, Todd pointed out, "is your new drive along the Rideau Canal" (Todd, 1903, p. 16). Replacing a jumble of industrial yards and private boat houses, the two-lane road along the canal's west bank subsequently became Queen Elizabeth's ceremonial Driveway.

It elevated the visual experience of entering an isolated lumber town, the civic symbolism of which was not lost on the Ottawa Improvement Commission (OIC), which had built the road. After starting work on a second parkway along the Ottawa River, in 1903 the OIC retained Todd to beautify the capital, and he recommended a system of park roads (Gordon, 2002).

Many of the roads first envisioned by Todd and the OIC were implemented in the Garden City/City Beautiful style, but only decades later, by other administrations, and after functionalist ideals had already eclipsed kinesthetic considerations of pleasure within city planning. Edward Bennett, a leading City Beautiful planner who collaborated with Burnham on the Chicago plan, elaborated the park road in his 1915 plan for Ottawa and Hull. The plan proposed to expand by 44 square miles a "system of parks, parkways and playgrounds" (Bennett, 1915, p. 130). It was shelved, however, as the nation focused on the war effort and rebuilding Parliament after a fire destroyed Centre Block. The quiescence of Bennett's parkway system was prolonged, moreover, by a shift away from City Beautiful thinking. The approach was heavily criticized for ignoring the poor housing conditions which often proliferated in the vicinity of grandiose civic centers (Gordon, 1998, p. 291; Hall, 2002, p. 191). During the 1920s, as a growing middle class could afford to take weekend pleasure drives outside the city in mass-produced cars, Ottawa turned instead to "scientific management" (Norton, 2008, p. 112), which prioritized efficiency and practical interventions for eliminating urban congestion. For example, Noulan Cauchon, a railway engineer and key actor in the expansion of municipal planning in Canada, elaborated a host of "City Scientific" interventions, like rounding street corners, street extensions, traffic regulation, rerouting and zoning by-laws (Gordon, 2008). Cauchon, with other first generation "planner-engineers," effectively paved the way for a new science of adapting urban space to the car (rather than the other way around). Still, even as this science became increasingly dominant in Canada, the park road persisted for three more decades as an object of planning.

Figure 1. Queen Elizabeth Driveway, Ottawa's first parkway, continues to manifest the elements of an urban park road in the 21st century.

The central features of the park road as a quasi-public, outwardly focused space designed for the automobile, extended into two important domains. The first domain was the production of Canada's inaugural superhighway, dedicated to the Queen who attended its opening on June 7, 1939. Following city parkways, the Queen Elizabeth Way (QEW) elaborated a number of technical innovations with which road engineers were experimenting across the continent to facilitate the movement of motorists (Jones, 2008; Norton, 2008). It featured a linear alignment, wrought iron guard rails, central medians, and the largest concatenation of lighting in the world. Yet, alongside this hard infrastructure, and partly as a disguise, were spruce and pine plantations, avenues of elm and maple, and other plant materials strategically organized to evoke a primeval landscape. The effect, Van Nostrand observed, was "a new highway landscape created in the image of the wilderness—a wilderness which, in the first instance, had been annihilated to make way for rural colonization" (1983, p. 9). A second domain in which the park road extended was one of the most comprehensively implemented master plans in Canadian history. The Gréber (1950) "Plan for the National Capital" embraced the parkways, boulevards, and expansive greenbelt proposed earlier by Todd, Bennett, and Cauchon. As a result, the national capital region today features 90 kilometers of scenic parkways "conceived to offer the public new ways to see and appreciate the landscape by car." As the National Capital Commission (NCC), successor to the OIC, elaborated: "These are not just roads. They are scenic gateways into a Capital experience. They link city and country and put people in touch with the Canadian landscape" (2011a, n.p.).

The second moment: Machine roads

While Gréber brought a belated parkway system to life, his plan acted as a fulcrum for an entirely different kind of road in the national capital region. The Gréber Plan, together with the reconstruction of the Queen Elizabeth Way (QEW) during the 1960s and 70s, illustrated a second critical planning moment of automobility. During this moment, three approaches to planning transformed the spatial basis of car driving. The first was an adapted set of Techno-Scientific City ideas. The early work of planner-engineers to reduce congestion by widening roads and rounding corners was expanded to create different classes of roads within a functionally segregated system of traffic. Such a system had already been established by New York and consummated in Los Angeles. It described an interconnected street hierarchy which prevented local traffic, filtered through slower collector roads and arterials, from mixing promiscuously with "through traffic" allowed to flow uninterrupted along restricted-access express roads (Brown, 2006). The second planning approach was Radiant City. Inspired by the Swiss architect, Le Corbusier, this approach, paradoxically, sought to obviate congestion in city centers by erecting uniform towers separated by open space and large expressways perched on viaducts. Le Corbusier's skyscrapers rarely found material expression. But they informed top-down schemes of urban renewal that cleared away poor neighborhoods for brutalist architecture. Finally, a third planning movement helped redefine the relationship between car and city, to which Hall (2002) aptly referred as "the City on the Highway." Pioneered by Robert Moses, New York's master builder, this approach refers to the power that large roads developed to shape patterns of urban growth.

A segregated system of traffic, top-down urban renewal, and the rise of car-oriented development all contributed to another critical planning moment, the machine road. Elements of a machine road first appeared in the freeway plans of the 1920s and 1930s, which proposed to adapt certain aspects of parkways such as their limited access (Brown, 2005, p. 9). Although these plans were delayed during the Great Depression, they were mobilized after WWII with Germany's autobahn project as a "point of reference" (Furness, 2010, p. 51). A well-oiled complex of interrelated actors, the machine road tends to dominate its surroundings. Unlike the park road, which manages but also cooperates with its context, trying to impress through every verge and vista its natural order upon the motorist, the machine road ploughs straight ahead, blinkered and efficient, while adjacent life turns its back. Today, Gréber is usually associated with Ottawa's beloved greenbelt, or, more critically, the odious manner in which his team of capital planners condemned a working class, largely French Canadian neighborhood, a "slum," which led to its being bulldozed (Jenkins, 1996). However, the centerpiece of the Gréber Plan, and Ottawa's new system of traffic, was an archetypal machine road. Gréber created a wide right of way by removing the east-west Canadian National Railway freight line, an elevated wall of industrial infrastructure which already had machinic qualities. The new road was designed to "ensure speedy through traffic, reduce present obstacles to north-south traffic movements, rehabilitate adjoining lands and relieve traffic congestion" (Gréber, 1950, p. 177). It took seven years to carve through the heart of Ottawa, and opened in 1965. Like the QEW, the expressway had a regal dedication. Similarly, Ottawa's Queensway subverted the traditional public meaning of Her Majesty's highway as open to all citizens regardless of their conveyance.

The Queensway adopted new technical standards which the QEW had developed since its pre-WWII opening in order to improve safety and efficiency. Unlike the Queensway, the QEW was not foisted above grade on top of an existing heavy rail line. Moreover, as already mentioned, its initial construction included outward-facing, park road pretenses towards its natural landscape. By the late 1940s, though, with the baby boom in motion and annual vehicle registrations quickly rising, the landscape surrounding the QEW was no longer "primeval" forest. It was a loose string of development that traced a "Golden Horseshoe." Coinciding with the growth that it had helped stimulate, the QEW earned a reputation as a congested death trap, with accidents commonly occurring at intersections and in front of the many private homes that clung onto the highway like barnacles. In response, the Department of Highways turned to its engineers. Over the 1950s, new service roads were constructed beside the existing highway in conjunction with grade-separated, cloverleaf interchanges to limit direct access to adjacent properties. These innovations did not necessarily undermine the QEW as a public, multi-use corridor. As Van Nostrand (1983) detailed, the single-family homes that first cropped along the first service roads faced the highway. Interchanges started to attract community facilities, such as Dixie Plaza, Canada's first shopping centre, whose posture also acknowledged the highway as a public space: "the overall effect was of a super-cornerstone located on a super-corner" (p. 14). Safety and efficiency on the super-corner, however, proved elusive. Parallel service roads and grade-separated interchanges, key "mediators" of machine roads (Latour, 2007), were extended along the QEW. Yet corresponding development, in a positive feedback loop, kept inviting cars and their corollaries, traffic and accidents, and then demanding more "improvements."

During the late 1960s and early 70s, the Department of Highways engineered a solution which settled the Queen's highways' utilitarian focus. It began widening the QEW from four to six, eight, and even ten lanes, a solution to which the Queensway also became committed quickly after its inception. More lanes meant bigger ramps and interchanges and, obviously, more car drivers. These were the "freeway conditions" with which the Queen's highways would cleanly cut through urban fabric and control car access. Freeways, however, were not as clean cut. They generated levels of traffic noise and tailpipe pollution that lessened the health of neighbors and reduced the value of adjacent property. Freeways also increased the amount of dead space which people had to hurry through at night, or the "structural holes" in urban space that can be dangerous particularly for women, older people, and the disabled (Sheller & Urry, 2000, p. 745). The machine road dominates its surrounding landscape, not by engaging it, but by disconnecting it from the view and concern of motorists. Bombarded by various kinds of pollution, buildings close to the QEW, increasingly commercial and industrial, turned to face the other direction (Van Nostrand, 1983). To be fair to Gréber, he envisioned an expressway for Ottawa with a civically inspiring view of the nation's capital as it came into focus through the windshield. He did not foresee the giant earth mounds and concrete barriers that later would be required to control the freeway's noise. Gréber also failed to foresee how explosive population growth, mixed with mass-produced cars and an ever expanding, publicly subsidized road network, would undermine a greenbelt meant to contain suburban sprawl.

Car-oriented development (COD) pivots on the extension of urban roads into forests and farmland to service low-density housing starts inaccessible by rail transit. COD sharply contrasts with the mode of tram-oriented development that coincided with park roads, marking a significant increase in the power of automobility to shape the built environment. Like many other Canadian cities, Ottawa built a thriving tram system which helped create dense, walkable hubs of housing and commerce around stations before the rise of COD. Demolishing streetcars, which the Gréber Plan depicted as ugly, depressing, unworthy of a capital, and, above all, "detrimental to traffic circulation" (1950, p. 127), was part of a multifaceted policy of decentralization. This policy began by removing urban rail infrastructure, including Ottawa's central train station, to the suburbs. It then proceeded by constructing peripheral office parks, expanding new "satellite" towns, and enrolling roads at the edge of the city into an interconnected urban grid (Gordon, 2001). Effectively, the city burst the compact streetcar suburb, spilling its contents horizontally across vast tracts of single-detached housing. Where earlier "developers would rarely build houses more than four blocks away from a streetcar line" (Hall, 2002, p. 304), under COD houses and retail were being fabricated in cheap, interstitial zones set apart from the freeways to which they were umbilically connected. Speedy development required standardized materials and mass production. Eastern Ontario's master builder, for example, the Minto Group, got its start in 1955 after four Ottawa brothers, Louis, Gilbert, Irving, and Lorry Greenberg, concocted an assembly line strategy to build houses on multiple sites simultaneously. By 1960, Minto boasted seven starts a day with a brisk, 72-day completion schedule. Five years later, after building more than 5,000 homes in Ottawa, Minto was "well on its way to becoming the region's largest private landlord" (Minto Group, 2011).

The third moment: New urban road

The machine road (freeway expansion, authoritarian renewal, and car-oriented development) has come under attack in contemporary planning (Doucet, 2007; Frumkin, Frank, & Jackson, 2004). Three alternative approaches have combined to force another critical planning moment of automobility. The first is new urbanism. Inspired by Jane Jacobs, who mobilized local opposition to Moses's plan to part her neighborhood for a freeway, it gained momentum from the mid-1960s. Related to transit-oriented development, new urbanism promotes dense, interconnected street patterns which support multiple, coexisting uses. The second approach, smart growth, criticizes COD, but also responds to anti-growth coalitions by seeking pragmatic ways to manage sprawl and coordinate jobs and housing, such as creating compact, mixed-use districts within car-oriented suburbs (Filion, 2003). Sustainable development, finally, has, since the Brundtland commission, inspired planning for environmental responsibility. These approaches blend together in recent conceptions of the city (Grant, 2009), and depict the new urban road. Like park roads, new urban roads attempt to insinuate themselves within their surroundings to produce a landscape conducive to civic interaction. But they trade romantic panoramas of nature and war monuments for the unpolished, unpredictable mixture of cultures and spaces that define urban coexistence. As Jacobs pointed out, advances in sanitation and public health have made urban density liveable: "Things have changed since the days when Ebenezer Howard looked at the slums of London and concluded that to save the people, city life must be abandoned" (Jacobs, 1961, p. 218). By not throwing the baby out with the bathwater the new urban road is a unique planning moment for automobility. It attempts to recover the spatial integration of the park road but also validate forms of mobility other than car driving that make a difference to urban life. New urban roads represent the good city as one in which strangers moving in multiple ways can accidentally encounter one another in a space where it is possible to interact.

Canada's capital city and most populous province, after moving away from top-down urban renewal schemes, have had only limited success with translating ideas associated with new urban roads into practice. It is not for lack of rhetorical support. In a complex planning arena shaped by federal laws, provincial policy, municipal regulations, and consumer choices, new urban roads enjoy nominal support in sub-national jurisdictions, including Ontario and British Columbia (Filion, Bunting, & Gertler, 2000; Grant, 2009). For example, the government of Ontario has passed legislation to demarcate greenbelts to help manage urban growth and promote intensification (Ontario Ministry of Infrastructure, 2006), while the City of Ottawa's (2003) master plan stresses the importance of coordinating land use and transportation policy to stimulate compact communities in which people can walk, cycle, and take public transit as well as drive. These plans picture the expansion of shared roads. A shared roadway moves people in cars from point A to point B, but is slow enough to also produce busy sidewalks, bike lanes, and dense, adjacent spaces of multiple uses that open onto it. In practice, new urban roads lie near the bottom of the traffic chain, often limited to the downtown core, and the new development leapfrogging over greenbelts tells a much different story. As Grant observed, while the ideals of new urbanism have caught on, "garage-front suburbs with cul-de-sacs of large and expensive homes remain common place," and "private communities" organized around the car continue to proliferate (2009, p. 15).

The Queen's highways to which these cul-de-sacs eventually drain in Ontario's vast commutershed, moreover, continue to grow. The Queensway may soon see more lanes to service Ottawa's outlying suburbs, whose population will outnumber those inside the greenbelt within the decade (Ontario Ministry of Transportation, 2010). The Queensway's only nod to new urbanism, again following the QEW, is carpool lanes to encourage passengering.

The success of new urban roads depends upon the ability of cities to limit the growth of machine roads and modify the heterogeneous associations on which this growth depends. The dominance of the machine road in the production of space lies in its capacity to persist as an inward-focused, utilitarian corridor monopolized by the "driver-car assemblage" (Dant, 2004). The machine road does not persist, however, as a static and path-dependent space, but as a dynamic, open-ended network where other social and technical actors—planner-engineers, cement rollers, suburban councilors, hydrocarbons, stimulus spending, the Ontario Municipal Board—constantly intervene and make modifications. Notwithstanding its dominance, the machine road has failed to fully efface earlier sedimentary layers such as the park road. As a salient example, the National Capital Commission's (NCC) parkway system offers a living testament to pre-machinic urban design, where car drivers can still access nature (at 60km/h) on exclusive roads with no commercial vehicles, limited signage and lighting, and carefully sculpted views of rivers and canals. Today the park road persists as a nostalgic landscape, and one that caters to (anti)urban rhythms unforeseen by early capital planners, such as extensive commuting to bedroom communities and the Ottawa International Airport. As the NCC admitted, "it is unlikely that such parkways will ever be built again. They should be cherished, therefore, as monuments to a bygone era in urban planning" (2011b, n.p.). Still, park roads and the old streetcar wards in which they reside, in spite of their enrollment into the "machinic complex" of automobility (Sheller & Urry, 2000, p. 738), point to an alternative way of integrating cars into the city. New urban roads, as outwardly connected streets that accommodate the mixing of different traffic participants, may benefit from expanding and transforming this earlier form of auto-space.

Conclusion

In three critical planning moments of automobility, I have argued, a number of novel associations between cars and the city created transformations in "constellations of mobility" (Cresswell, 2010). In the first moment, car driving was imagined as an optimal mode of connecting with nature in the city, albeit a stylized nature crafted for the motorist's gaze. Such outwardly focused pretenses were abandoned during the second moment. The machine road, like the park road, caters exclusively to the car driver, but with an overriding concern for power, circulation and speed. Finally, in a third critical moment, the impermeability of the machine road is contested. Furthermore, the new urban road challenges exclusive road networks in general that privilege the uninterrupted displacement of cars. Instead of rigid segregation it promotes the interconnection of multiple mobilities, including non-motorized forms of automobility such as cycling (Furness, 2010). A crucial question moving forward is whether or not North American cities dominated by cars can capitalize on this third moment and reassemble the city. Like the park road and the machine road, the new urban road is not simply a set of technical or politically neutral precepts, but a way of organizing a civic practice with consequences for urban civility and solidarity. What makes the new urban road distinctive, however, is that it imagines the good city as one that accommodates difference through automobility. Implementing this vision will require the transformation of hard infrastructures that suggest roads that are monopolized by cars constitute natural or inevitable features of the urban environment.

Encouraging signs of growth in new urban roads are emanating from all over the region which pioneered Canadian automobility. Car sharing, park-and-ride programs, reintensification policies and, significantly, light rail in Ottawa appear to be gaining appeal. An important test facing Canada's capital is whether it can stimulate higher levels of cycling. As in the case of urban Canada at large, cycling rates remain stubbornly low in Ottawa, around two percent of person-trips, with safety concerns about cycling alongside cars posing a major obstacle (City of Ottawa, 2008). But in a notable shift, the National Capital Commission has begun to rethink the extensive 175 kilometer network of multi-use pathways that it maintains in the capital region in order to make it more practical. Ottawa is frequently pictured as a cycling-friendly city largely because of these scenic lanes, although in essence they provide "park roads" for recreational use with relatively few linkages for utility cyclists. But in the fall of 2010, the NCC, along with local officials, travelled to northern European cities on a fact-finding mission to determine how Ottawa can "invite" cyclists into the city (Gehl, 2010). The planners returned intending to "Copenhagenize" Canada's capital and turn cycling into a viable form of automobility.

References

Amin, A. (2006). The good city. *Urban Studies, 43*(5), 1009–1023.
Amin, A., & Thrift, N. (2002). *Cities: Reimagining the urban*. Malden, MA: Blackwell.
Beckmann, J. (2001). Automobility—A social problem and theoretical concept. *Environment and Planning D: Society and Space, 19*(5), 593–607.

Bennett, E. (1915). *Report of the Federal Plan Commission on a general plan for the cities of Ottawa and Hull.* Ottawa, Ontario, Canada: Federal Plan Commission. Retrieved from https://qshare.queensu.ca/Users01/gordond/planningcanadacapital/bennett1915/index.htm

Best, A. (2006). *Fast cars, cool rides: The accelerating world of youth and their cars.* New York, NY: New York University Press.

Brown, J. (2005). A tale of two visions: Harland Bartholomew, Robert Moses, and the development of the American freeway. *Journal of Planning History, 4*(1), 3-32.

Brown, J. (2006). From traffic regulation to limited ways: The effort to build a science of transportation planning. *Journal of Planning History, 5*(1), 3-34.

Brugge, D., Durant, J. L., & Rioux, C. (2007). Near-highway pollutants in motor vehicle exhaust: A review of epidemiologic evidence of cardiac and pulmonary health risks. *Environmental Health, 6*(1). Retrieved from http://www.ncbi.nlm.nih.gov/pmc/articles/PMC1971259/

Bull, M. (2004). Automobility and the power of sound. *Theory, Culture & Society, 21*(5), 243-259.

Bullard, R. D., Johnson, G. S., & Torres, A. O. (2004). *Highway robbery: Transportation racism & new routes to equity.* Cambridge, MA: South End Press.

Callon, M. (1989). Society in the making: The study of technology as a tool for sociological analysis. In W. E. Bijker, T. P. Hughes, & T. Pinch (Eds.), *The social construction of technological systems: New directions in the sociology and history of technology* (pp. 83-106). Cambridge, MA: MIT Press.

City of Ottawa. (2003). *Ottawa 20/20: Ottawa's Official Plan.* Retrieved from http://ottawa.ca/city_services/planningzoning/2020/transpo/toc_en.shtml

City of Ottawa. (2008). *Ottawa Cycling Plan.* Retrieved from http://www.ottawa.ca/residents/onthemove/cycling/ottawa_cycling_plan_en.html

Conley, J., & McLaren, A. T. (Eds.). (2009). *Car troubles: Critical studies of automobility and auto-mobility.* Burlington, VT: Ashgate.

Cresswell, T. (2010). Towards a politics of mobility. *Environment and Planning D: Society and Space, 28*(1), 17-31.

Dant, T. (2004). The driver-car. *Theory, Culture & Society, 21*(4-5), 61-79.

Dean, J. (1996). *Solidarity of strangers: Feminism after identity politics.* Berkley, CA: University of California Press.

Dennis, K., & Urry, J. (2009). *After the car.* Malden, MA: Polity Press.

Doucet, C. (2007). *Urban meltdown: Cities, climate change and politics as usual.* Gabriola Island, British Columbia, Canada: New Society Publishers.

Eckstein, B. (2003). Making space: Stories in the practice of planning. In B. Eckstein & J. A. Throgmorton (Eds.), *Story and sustainability: Planning, practice, and possibility for American cities* (pp. 13-38). Cambridge, MA: MIT Press.

Farías, I., & Bender, T. (2010). (Eds.). *Urban assemblages: How actor-network theory changes urban studies.* London, England: Routledge.

Filion, P. (2003). Towards smart growth?: The difficult implementation of alternatives to urban dispersion. *Canadian Journal of Urban Research, 12*(1), 48-70.

Filion, P., Bunting, T., & Gertler, L. (2000). Cities and transition: Changing patterns of urban growth and form in Canada. In T. Bunting & P. Filion (Eds.), *Canadian cities in transition: The twenty-first century* (pp. 1-25). Toronto, Ontario, Canada: Oxford University Press.

Flyvbjerg, B. (1998). *Rationality and power: Democracy in practice.* Chicago, IL: University of Chicago Press.

Frumkin, H., Frank, L., & Jackson, R. (2004). *Urban sprawl and public health: Designing, planning, and building for healthy communities.* Washington, DC: Island Press.

Furness, Z. (2010). *One less car: Bicycling and the politics of automobility*. Philadelphia, PA: Temple University Press.

Galston, W. (1991). *Liberal purposes: Goods, virtues, and diversity in the liberal state*. Cambridge, England: Cambridge University Press.

Gehl, J. (2010). *Cities for people*. Washington, DC: Island Press.

Gordon, D. L. A. (1998). A City Beautiful plan for Canada's capital: Edward Bennett and the 1915 plan for Ottawa and Hull. *Planning Perspectives, 13*(3), 275-300.

Gordon, D. L. A. (2001). Weaving a modern plan for Canada's capital: Jacques Gréber and the 1950 plan for the National Capital Region. *Urban History Review*, 29(2), 43-61.

Gordon, D. L. A. (2002). Frederick G. Todd and the origins of the park system in Canada's capital. *Journal of Planning History, 1*(1), 29-57.

Gordon, D. L. A. (2008). 'Agitating people's brains': Noulan Cauchon and the City Scientific in Canada's capital. *Planning Perspectives, 23*(3), 349-379.

Grant, J. (2009). Theory and practice in planning the suburbs: Challenges to implementing new urbanism, smart growth, and sustainability principles. *Planning Theory & Practice*, 10(1), 11-33.

Gréber, J. (1950). *Plan for the National Capital: General report submitted to the National Capital Planning Committee*. Ottawa, Ontario, Canada: National Capital Planning Service. Retrieved from https://qshare.queensu.ca/Users01/gordond/planningcanadascapital/greber1950/index.htm

Hall, P. (2002). *Cities of tomorrow: An intellectual history of urban planning and design in the twentieth century* (3rd ed.). Oxford, England: Blackwell.

Hommels, A. (2008). *Unbuilding cities: Obduracy in urban socio-technical change*. Cambridge, MA: MIT Press.

Hubbard, P., & Lilley, K. (2004). Pacemaking the modern city: The urban politics of speed and slowness. *Environment and Planning D: Society and Space, 22*(2), 273-294.

Jacobs, J. (1961). *The death and life of great American cities*. New York, NY: Random House.

Jenkins, P. (1996). *An acre of time*. Toronto, Ontario, Canada: Macfarlane Walter & Ross.

Jensen, O. B. (2007). Culture stories: Understanding cultural urban branding. *Planning Theory*, 6(3), 211-36.

Jones, D. W. (2008). *Mass motorization and mass transit: An American history and policy analysis*. Bloomington, IN: Indiana University Press.

Kymlicka, W. (2002). *Contemporary political philosophy: An introduction*. Oxford, England: Oxford University Press.

Kymlicka, W., & Norman, W. (Eds.). (2000). *Citizenship in diverse societies*. Oxford, England: Oxford University Press.

Latour, B. (1997). Trains of thought: Piaget, formalism, and the fifth dimension. *Common Knowledge, 6*(3), 170-191.

Latour B. (2007). *Reassembling the social: An introduction to actor-network-theory*. New York, NY: Oxford University Press.

Law, J. (1999). After ANT: Complexity, naming and topology. In J. Law & J. Hassard (Eds.), *Actor network theory and after* (pp. 1-14). Oxford, England: Blackwell.

Lefebvre, H. (1991). The production of space (D. Nicholson-Smith, Trans.). Oxford, England: Blackwell.

Lupton, D. (1999). Monsters in metal cocoons: 'Road rage' and cyborg bodies. *Body and Society, 5*(1), 57-72.

Marshall, T. H. (1950). *Citizenship and social class and other essays*. Cambridge, England: Cambridge University Press.

Minto Group. (2011). *An inspired beginning: History of Minto.* Retrieved from http://www.minto.com/history_of_minto.html

Murray, L. (2008). Motherhood, risk and everyday mobilities. In T. P. Uteng & T. Cresswell (Eds.), *Gendered mobilities* (pp. 47–63). Aldershot, England: Ashgate.

National Capital Commission (NCC). (2011a). *Capital pathways and parkways.* Ottawa, Ontario, Canada: National Capital Commission. Retrieved from http://www.canadascapital.gc.ca/

National Capital Commission (NCC). (2011b). *Managing parkways.* Ottawa, Ontario, Canada: National Capital Commission. Retrieved from http://www.canadascapital.gc.ca/bins/ncc_web_content_page.asp?cid=16300-20448&lang=1

Norton, P. D. (2008). *Fighting traffic: The dawn of the motor age in the American city.* Cambridge, MA: MIT Press.

Olmsted, F. L. (1997). *Civilizing American cities: Writings on city landscapes* (S. B. Sutton, Ed.). New York, NY: Da Capo Press.

Ontario Ministry of Infrastructure. (2006). *Places to grow: Growth plan for the Greater Golden Horseshoe.* Retrieved from https://www.placestogrow.ca/index.php?option=com_content&task=view&id=9&Itemid=14&lang=eng

Ontario Ministry of Transportation. (2010). *Ottawa Queensway preliminary design and environmental assessment study.* Retrieved from http://www.mto.gov.on.ca/english/engineering/417ea/

Packer, J. (2008). *Mobility without mayhem: Safety, cars, and citizenship.* Durham, NC: Duke University Press.

Parusel, S., & McLaren, A. T. (2010). Cars before kids: Automobility and the illusion of school traffic safety. *Canadian Review of Sociology, 47*(3), 129–147.

Paterson, M. (2007). *Automobile politics: Ecology and cultural political economy.* Cambridge, England: Cambridge University Press.

Preston, J., & Raje, F. (2007). Accessibility, mobility and transport-related social exclusion. *Journal of Transport Geography, 15*(3), 151–160.

Rajan, S. C. (2006). Automobility and the liberal disposition. *The Sociological Review, 54*(1), 113–129.

Routledge, P. (2010). Introduction: Cities, justice and conflict. *Urban Studies, 47*(6), 1165–1177.

Scott, J. (1998). *Seeing like a state: How certain schemes to improve the human condition have failed.* New Haven, CT: Yale University Press.

Sheller, M. (2004). Automotive emotions: Feeling the car. *Theory, Culture & Society, 21*(4-5), 221–242.

Sheller, M., & Urry, J. (2000). The city and the car. *International Journal of Urban and Regional Research, 24*(4), 737–757.

Short, J. R., & Pinet-Peralta, L. M. (2010). No accident: Traffic and pedestrians in the modern city. *Mobilities, 5*(1), 41–59.

Soron, D. (2009). Driven to drive: Cars and the problem of 'compulsory consumption.' In J. Conley & A. T. McLaren, (Eds.), *Car troubles: Critical studies of automobility and automobility* (pp. 181–196). Burlington, VT: Ashgate.

Staudenmaier, J. (1985). *Technology's storytellers: Reweaving the human fabric.* Cambridge, MA: MIT Press.

Taylor, C. (2004). *Modern social imaginaries.* Durham, NC: Duke University Press.

Thrift, N. (2004). Driving in the city. *Theory, Culture & Society, 21*, 41-60.

Todd, F. G. (1903). *Preliminary report to the Ottawa Improvement Commission.* Ottawa, Ontario, Canada: Ottawa Improvement Commission. Retrieved from https://qshare.queensu.ca/Users01/gordond/planningcanadascapital/todd1903/Title.htm

Transport Canada. (2010). *Road transportation.* Retrieved from http://www.tc.gc.ca/eng/road-menu.htm

Turcotte, M. (2008). Dependence on cars in urban neighbourhoods. *Canadian Social Trends, 85,* 20-30.

Urry, J. (2004). The 'system' of automobility. *Theory, Culture & Society, 21*(4-5), 25-39.

Urry, J. (2006). Inhabiting the car. *Sociological Review, 54*(1), 17-31.

Vanderbilt, T. (2009). *Traffic: Why we drive the way we do (and what it says about us).* Toronto, Ontario, Canada: Random House.

Van Nostrand, J. C. (1983). The Queen Elizabeth Way: Public utility versus public space. *Urban History Review, 12*(2), 1-23.

Young, I. M. (1990). *Justice and the politics of difference.* Princeton, NJ: Princeton University Press.

6

Selling the World: Airline Advertisements and the Promotion of International Aeromobility in *National Geographic*, 1964–2004

Lucy Budd

National Geographic has been a part of my life for almost as long as I can recall. Both my father and my late grandfather subscribed to it and, as a young child, I can vividly remember looking through the magazine's glossy pages and imagining myself visiting the strange and wonderful places that were depicted therein. The publication proved to be an indispensable aid to various pieces of high school homework, and it was a familiar reminder of home when I went away to university. Five years ago, while helping my grandparents to move house, I came across my grandfather's collection of old issues of *National Geographic* and my interest in the title was rekindled. Despite their age (some dated back to the late 1940s), each issue had been perfectly preserved and the contents offered a tantalizing glimpse into the past.

In addition to featuring lavishly illustrated articles on recent geographic explorations and anthropological encounters, the magazines also contained a number of advertisements that promoted the consumption of luxury consumer goods, including various pieces of optical equipment and different models of automobile. It was, however, a half-page advertisement for South African Airways that caught my eye. "Come where the giraffes, klipspringers, and tsessebes play," it implored, and, quite spontaneously, I found myself imagining that I was standing in the African bush with the roar of lions in the distance and the warmth of the sun on my face. Returning reluctantly to reality, yet curious about the advertisement's ability to affect my thoughts and mentally transport me to a country that I had never visited, I turned the page and found another similarly enticing proposition. "Wherever the land of your dreams—no matter how seemingly distant—it is only a few hours away on Lufthansa," it read.

By alerting readers to the existence of a world beyond the confines of their everyday environment and emphasizing the relative ease with which exotic destinations could be accessed and consumed by air, both advertisements were seeking to generate a desire for international travel. In so doing, they actively espoused a particular form of post-war international (aero)mobility that was predicated on speed and unfettered access to worldwide networks of commercial air services. In addition to stirring my imagination, the two advertisements also caused me to reflect on how the co-evolution of multiple technologies (including the airplane but also print, advertising media, and worldwide postal distribution) had simultaneously enabled and stimulated demand for a new type of mobility that transcended the boundaries of the everyday and metaphorically brought geographically distant people and places closer together in time and space. Crucially, and despite the existence of a substantial body of literature which critically analyzes the articles and the visual images that have been published in *National Geographic* (see Lutz & Collins, 1991; Rothenberg, 2007), no equivalent research has examined the nature or content of the advertisements that were published alongside them. This is unfortunate, as my encounter with the South African Airways and Lufthansa examples led me to believe that advertisements are important cultural artifacts that can reveal much about changing societal attitudes and practices of mobility and consumption.

In recognition of the absence of any academic consideration of the advertisements within *National Geographic*, the present research seeks to identify the strategies that were used by commercial airlines in an attempt to "sell the world" to potential passengers and generate a desire for international travel and wanderlust among the magazine's estimated 38 million readers. The chapter begins with a discussion of aviation in American culture and the role of *National Geographic* in disseminating news of aeronautical achievement and shaping public perceptions of the world beyond the U.S. border. This is followed by a description of the method that was employed. Then the findings are presented and their implications for furthering academic understandings of past practices of aeromobility are discussed.

Commercial Aviation and *National Geographic*

To mark the centenary of Orville and Wilbur Wright's first successful heavier-than-air powered flights on the windswept sand dunes of Kill Devil Hills, near Kitty Hawk, North Carolina, the December 2003 issue of *National Geographic* contained an article on the future of flying. The cover image, framed by the magazine's instantly recognizable, vivid yellow border, featured a dramatic close-up photograph of the nose of one of the United States Air Force's new supersonic stealth fighters, the F/A-22 Raptor. Inside, and accompanied by vivid, full-colour, air-to-air photographs of close formation flying, backlit images of high performance turbofan engines, and detailed descriptions of computerized flight decks, the article reflected on 100 years of aeronautical achievement and postulated a future in which hypersonic aircraft would be able to reach any point on the earth's surface in under four hours (Klesius, 2003). The article spoke to a technologically sophisticated, globalized, and highly interconnected world in which aviation had overcome the tyranny of distance, rendering geographically remote countries effectively near neighbors and enabling affluent members of global society to participate in worldwide networks of trade, travel, and commerce.

Far from being a one-off, the article represented the latest in a long line of features that the magazine had published on the subject of flight. As part of the National Geographic Society's mission to promote "the increase and diffusion of geographical knowledge," over 115 individual articles concerning different aspects of military and commercial aeronautics were published in its magazine between 1918 and 2010. The earliest articles considered aviation's role in World War One (see De Sieyes, 1918; Grosvenor, 1918; Tulasne, 1918), the industry's strategic importance to U.S. national defense and future economic development (Mitchell, 1921), and air travel's role in "opening up" remote regions of the world to geographic exploration and anthropological discovery (see, for example, Cobham, 1928; Dargue, 1927; Grosvenor, 1924; Van Zandt, 1925; Wilson, 1926). Later features addressed the development of post-World War Two aviation infrastructure (Colton, 1948) and described how the introduction of increasingly sophisticated aeronautical technologies, including autopilots, fly-by-wire controls, and instrument landing systems, would improve the safety and efficiency of the air transport system (Long, 1977; Klesius, 2003).

If one subscribes to the arguments advanced by aviation historians, Joseph Corn (1983), Joe Christy (1987), Tom Crouch (2003), and Robert Wohl (2005), aviation performs an important role in American society and occupies a particular place in the American national psyche. Certainly, in a little over a hundred years between the birth of the modern aerial age and today, air travel has emerged as *the* normal and dominant mode of long-distance mobility for a significant (and growing) section of American society, and flying has become a routine activity for many. Every day, tens of thousands of U.S. citizens travel to airports, buckle themselves into airline seats, fly through the troposphere at close to the speed of sound, and deplane at their destinations a few minutes or hours later. They do so not because they actively enjoy the experience of flight (although some undoubtedly do), but because their personal and/or professional lives demand the routine performance of high levels of long-distance mobility. At any given time, it has been estimated that as many as 300,000 people and 6,200 commercial flights will be airborne above the United States (Klesius, 2003), and such habitual volumes of aerial mobility have arguably become one of the defining features of contemporary American (and increasingly global) society.

Modern commercial airplanes and air traffic control systems enable considerable numbers of people to routinely undertake long-distance journeys (that they believe to be necessary) to distant places (that they perceive to be desirable) to pursue particular experiences or encounters that are not available at home, both safely and also more quickly and more cheaply than alternative transport modes. Yet while much has been written about the changing spatialities of air service provision and the growth in global air travel that modern aeronautical technologies have effected (see Graham, 1995; Hanlon, 1996), it is only relatively recently that the cultural and embodied dimensions of human aeromobility have begun to be systematically explored (Adey, 2010; Budd, 2011; Millward, 2008; Rust, 2009).

The Commercial Aerial Age

The origins of commercial flying in the United States can be traced back to the Wright brothers' pioneering, heavier-than-air flights in 1903 and the first domestic air mail services that were inaugurated by the U.S. Post Office and the U.S. Air Service in 1918 (Christy, 1987). Despite only seeing limited service during World War One, the potential for aircraft to be subsequently employed on peacetime civilian operations had been recognized, and immediately after the conflict ended American aircraft manufacturers began producing more sophisticated and reliable aircraft for commercial use. The development of these new aircraft enabled pioneering U.S. airlines to begin offering regular, scheduled services for paying passengers. A period of regulatory and operational reform during the early 1920s resulted in formation of specific national regulations for civil aviation. This framework enabled a growing band of American entrepreneurs to begin exploiting aviation's commercial potential by conveying growing volumes of passengers, mail, and freight. The media coverage that accompanied Charles Lindbergh's successful, transatlantic flight between New York and Paris in May 1927 stimulated further interest in the scientific discipline of aeronautics and ensured that aviation entered mainstream American public consciousness.

During the 1930s and 1940s, millions of dollars were invested in developing new aircraft and aero engines that could outperform all existing machines and, by the end of the Second World War, American aerospace companies, including Lockheed, Pratt and Whitney, Douglas, and Boeing, were well placed to begin construction of a new generation of post-war, commercial aircraft. Although the world's first jet-powered, commercial aircraft had been designed in Britain, structural problems with the airframe ensured that American manufacturers quickly gained the ascendancy and ultimately dominated the market for post-war, jet-powered passenger aircraft. With their sleek, aerodynamic fuselages, superior speed and range, and sophisticated technology, the new American-built "jetliners" of the late 1950s and early 1960s represented all that was exciting and progressive about modernity. The new airframes, including Boeing's 707 and Douglas's DC-8, could fly more people further, faster, longer, higher, and more economically than the piston- and propeller-engined aircraft they replaced. Journeys that had once taken the best part of a day to complete could now be accomplished in a matter of a few minutes or hours. The British-designed Comet, for example, reduced the flight time from London to Johannesburg from 32½ hours to 18, enabled Singapore to be reached in 25 hours rather than 2½ days, and cut the flight time from London to Tokyo from 86 to 33¼ hours (Hensser, 1953). Continued innovations in aircraft design and performance during the 1960s and 1970s progressively reduced the monetary cost of air fares and stimulated a new vogue for international travel and long-distance aeromobility.

These wide-ranging social and technological changes were predicated on a number of diverse yet interlocking factors. For aviation historian, Roger Bilstein (1984, p. 232), the growth of post-war civil aviation in the United States reflected "a host of changes and social effects that had been gathering momentum" for a number of years. Increasing affluence, growing social mobility, and enhanced technological capabilities in the fields of material sciences, aerodynamics, microelectronics, and propulsion, enabled growing numbers of American citizens to take to the air. By the time the Boeing Company launched its iconic 747 "Jumbo Jet" in 1969, commercial airlines were already carrying 17.1 million people a year, a figure that had increased from 3.1 million a decade earlier. In a spirit of post-war optimism, U.S. politicians and social commentators alike asserted that the provision of regular, non-stop, international airline services to and from the United States would not only promote global peace and understanding but also open up foreign markets to American overseas trade and investment (Crouch, 2003).

In addition to creating new opportunities for international business and communication, post-war passenger aviation also transformed U.S. citizens' leisure and vacation habits. Whereas well-heeled American tourists of the 1920s and 1930s traditionally spent their summer holiday in Atlantic coast resorts such as Cape Cod or Atlantic City, the rapid growth of air travel from the early 1960s onwards meant that far-flung and ever more exotic destinations were suddenly within reach. Airline advertisements were instrumental in alerting Americans to these new travel opportunities, and the marketing rhetoric the campaigns employed resonated with, and reinforced, the developing worldview of a new generation of young, increasingly affluent, and fashion conscious "jet set" travelers. Jet flight was promoted as offering a portal into a streamlined, exciting, and fashionable future in which the whole world was a playground, and the sun-kissed beaches of the Caribbean, the fashion houses of Europe, or the plains of Africa were a mere few hours' flying time away.

Reconstructing Past Mobility Regimes From the Pages of *National Geographic*

Within the last decade, in particular, scholars from across the social sciences have become increasingly attuned to the myriad ways in which people move, and to the socio-cultural, economic, and environmental significance of these different forms of mobility. Taking their cue from the seminal work of John Urry (2000, 2007), geographers, anthropologists, sociologists, psychologists, and cultural historians, among others, have employed innovative research techniques to examine the mobility patterns and spatio-temporal practices of everything from routine daily commutes to student gap year travel, international business trips, and "once in a lifetime" adventure tourism. Some of this work has been motivated by a desire to better understand the practical decision-making of business travelers and tourists, whereas other research has sought to examine the personal and/or embodied dimensions different modes of transportation engender (see, for example, Bissell, 2008, 2009; Edensor, 2003).

In the context of commercial aviation, such approaches have led to examinations of the socio-cultural and "affective" experiences of air travel and have prompted considerations of the extent to which human language, travel choices, diets, working practices, and experiences of becoming and being mobile have been shaped by developments in aeronautical technology and commercial aviation practice (see, for example, Adey, 2010; Budd, 2010, 2011; Gottdiener, 2001; Pascoe, 2001). Such accounts serve as effective counterpoints to other literature which has merely commented on the technological innovations that resulted in aviation evolving from a specialized activity, pursued only by a few wealthy and/or foolhardy individuals, into a multibillion dollar enterprise that facilitates the global mobility of over 2.4 billion passengers a year.

In addition to altering the trajectory of conventional transport research, the "mobilities turn" has also been instrumental in stimulating innovations in empirical practice. A growing number of academic studies now employ historical artifacts, including personal travel diaries, postcards, timetables, travel brochures, and other tourist ephemera, to help reconstruct past mobility regimes and further our understanding of past practices of movement (Walsh, 1990; Watts, 2004). Advertisements, in particular, have become an increasingly popular object of academic inquiry, as their marketing rhetoric and visual presentation provide "a window on to the landscape of our social culture" (Lyth, 2009, p. 2). Far from being mere transient promotional tools, advertisements speak of changing socio-cultural fashions, trends, and tastes, and their study offers valuable insights into how particular brands or industrial sectors, including travel and transportation companies, present themselves to the public and evolve over time. This awareness has led to a panoply of studies which have examined the design histories and corporate publicity strategies of interwar, British railway companies (Harrington, 2004; Hewitt, 2000; Watts, 2004), Atlantic steamship lines (Swinglehurst, 1982), and American bus companies (Walsh, 1990). However, despite an emerging literature on the role advertisements play in the promotion of particular forms of maritime and surface transport mobility, and an established body of popular or enthusiast texts which examine the graphic design histories of individual airlines and aerospace companies (see, for example, Cruddas, 2008; London, 2007; Lovegrove, 2000; Remmele, 2004; Szurovy, 2002), little academic research has examined how airlines use printed advertising media to generate a desire for travel and wanderlust among potential passengers.

National Geographic was selected as the source material for this analysis owing to the availability of examples and also to the publication's status in, and importance to, American popular culture. Launched in 1888 as a sporadic scientific journal, the publication rapidly evolved into a monthly magazine that disseminated news of geographic exploration, technological development, and cultural anthropology to the Society's growing membership. Many of the articles conveyed details of daring, Society-funded expeditions and informed legions of "armchair travelers" about the earth, its seas and sky, and outer space. By the 1950s, the magazine's format was firmly established and its archetypal "white, Christian, middle-class, small-town," American readers (Rothenberg, 2007, p. 2), were receiving a regular diet of carefully selected and scripted information on the world as the Society's correspondents and editors saw it. The magazine's core editorial objectives, its high production values (which, since the mid-1960s, included the extensive use of color printing; glossy, heavyweight paper; and the provision of pull-out maps and posters), and its iconic title page framed in vivid yellow have made *National Geographic* one of the world's most popular and visible expressions of geographic knowledge. From a print run of 750,000 copies during the early 1920s, the magazine's circulation grew steadily during the twentieth century to the point where approximately eight million copies are now published every month. The Society estimates that each issue is read by around 38 million people worldwide and the title is published in 34 languages (National Geographic, 2011).

Owing to the title's extensive circulation and readership, Lutz and Collins (1991, p. 1) opined that, during the twentieth century, *National Geographic* was one of the "primary means by which people in the United States receive[d] information and images of the world outside their borders." The articles and images the magazine contained undoubtedly helped foster awareness of the world beyond the U.S. border. However, critics have suggested that the selection and (re)presentation of certain "foreign" cultures and civilizations served to create a distinct American identity based on notions of U.S. "civil and technological superiority" (Rothenberg, 2007, p. 5; see also Abramson, 2010; Bryan, 1987; Jansson, 2003; Pauly, 1979; Tuason, 1999). Arguably, this discourse of U.S. capability and "superiority" was created and reinforced not only by the articles but also by the advertisements each issue contained.

A cursory glance through any post-1945 issue of *National Geographic* reveals the presence of multiple advertisements for high-end luxury goods and services, including designer watches, automobiles, high specification photographic and optical equipment, and discretionary travel. The latter group includes advertisements for overseas tourist agencies, airline operators, and international hotel chains, all of which aimed to generate a desire for wanderlust by emphasizing the ease, affordability, and rapidity with which exotic (and, therefore, exciting) destinations could be accessed and consumed. In order to obtain data both on the frequency with which airline advertisements appeared and also the nature of the marketing message they employed, I performed a detailed content analysis of all the airline advertisements that appeared in all 492 issues of *National Geographic* that were published between January 1964 and December 2004. Content analysis was selected as it enables "the objective, systematic, and quantitative description" of the manifest content of communication to be undertaken, and allows the manifest content of different types of media to be objectively evaluated and recorded (Berelson, 1952, p. 18). A preliminary scoping study of 50 randomly selected examples revealed that every advertisement could be classified into one of three broad themes according to whether its primary selling point was an airline's geographic reach, its customer service, or its technological attributes.

Advertisements that emphasized a carrier's worldwide scale, scope, and global connectivity, or which conveyed details of a new route launch were classified as geographical attributes. Those which promoted various aspects of an airline's product, including affordability/value for money, seat comfort, leg room, the quality of in-flight entertainment and in-flight food, and the professionalism and attentiveness of cabin crew and pilots, were described as primarily promoting service characteristics, while those that made reference to particular types of aircraft or aeronautical technologies, such as on-board weather radar or automatic instrument landing systems, were classified as marketing an airline's technological attributes.

Selling the World: Principal Findings and Discussion

1181 advertisements, collectively promoting 56 different airlines and one global airline alliance, were identified and coded. Of the total, 61% were for European carriers, and North American operators only accounted for 17% of all the advertisements examined. Lufthansa, the German national carrier, and Air France were the most prolific advertisers, placing 205 and 126 advertisements respectively during the forty-year period under examination. The U.S. airline with the most advertisements, the now-defunct Trans World Airlines (TWA), placed only 70. The dominance of European operators suggests that the type of aeromobility that was being promoted through the pages of *National Geographic* was very specific and geared towards generating demand for U.S. consumers to undertake international, as opposed to domestic U.S. travel. An indication of the variety of airline advertisements that were published is provided in Figure 1.

Figure 1: Examples of airline advertisements in National Geographic.
Photograph: Author

Of the three coding categories, geographical attributes were the most frequently employed method by which airlines attempted to "sell the world" to consumers, accounting for 47% of all the advertisements that were identified. Williams (1998) argued that historical patterns of global tourist and business travel mobilities are, to a significant degree, shaped by how individual places are perceived by different groups of consumers and how these perceptions change over time in response to new geopolitical, economic, and environmental conditions. While images of distant places have long been used to generate a desire for mobility and travel (see Watts, 2004), post-war airlines arguably took the creative promotion of tourist spaces to a new level. In order to stimulate consumer demand for flight, airlines had to convince potential customers that flying was not only safe and comfortable but also that it offered unrivalled opportunities to further professional development and personal fulfillment. The processes involved in packaging and then selling the world to American travelers required the selective (and often highly politicized) transformation of ordinary and everyday places into exciting travel destinations. Many of the airline advertisements in *National Geographic* were unashamedly escapist in tone and used lavish color photographs and lengthy textual descriptions to emphasize the difference, exoticism, timelessness, romanticism, and/or cultural attraction of the place(s) they were seeking to promote.

As Urry (1990) has explained, the role of vision and, in particular, the "tourist gaze," is intrinsically bound up with notions of difference. Foreign places have to be seen to be sufficiently different so as to render them interesting (and thus worthy of visit) but simultaneously not so extreme as to render them dangerous or unsettling. In order to reassure potential passengers that the airlines took consumer concerns about the potential for culture shock seriously, Pan American published a traveler's companion book for prospective passengers. This "New Horizons World Guide" contained information on where and when to travel, what to pack, how to behave, what to see and do in different locations, and how to successfully navigate one's way through international airports. The guide professed to contain all the information that any potential traveler might want (or need) to know about the idiosyncrasies of global aviation and foreign customs but was too afraid to ask.

As one of the premium travel brands of the 1960s, Pan Am evidently took its educational responsibilities seriously. An advertisement from the late 1960s, for example, featured a large, color photograph of a relaxed and carefree, young, (White) American couple strolling along the bank of the River Seine in Paris. The advertisement was accompanied by the reassuring message:

> The Martins had never been to Europe before. That's why they came to us....We told them how to get a passport. What to pack. How to plan an itinerary. How to clear customs. Where to look for bargains on the left bank. Where to be seen on the Via Veneto. How to tell if a restaurant's expensive without walking in...

In short, the advertising copy sought to contain everything that any self-respecting, yet perhaps inexperienced, American tourist might want to know in order to be able to relax and enjoy their trip. In addition to acting as a "traveler's friend," Pan Am's advertisements also sought to stimulate a desire for wanderlust by emphasizing the number of different destinations served by their aircraft. Many advertisements from the mid-1960s featured specific destinations and provided glossy photographs and descriptions of the delights that could be found there in an attempt to entice people into the air. In 1966 the airline boasted,

> We fly to more Caribbean favorites than anybody...15 in all, and every one different. Head for Barbados or Antigua for British accents. Mark down Martinique and Guadeloupe for French flavor. Go Latin in Puerto Rico. Or try our Dutch treats—Curacao and Aruba. Just pick your sun spot. Then call your Pan Am travel agent. Or call us. And fly away with the best there is in the world.

The ways in which destinations are visualized by travelers was (and remains) highly subjective and politicized (Williams, 1998). Places that were once favored and considered desirable destinations to see and be seen in can rapidly be marginalized in favor of newer, "up and coming" resorts. As air travel became an increasingly routine activity, the nature of the advertisements changed and airlines began promoting ever more distant destinations. The trend towards the consumption of increasingly exotic destinations was evident in displays of geographical and corporate "one-upmanship" in which airlines engaged in competition with their rivals by advertising the inauguration of ever more unusual, preferably "undiscovered," and therefore "exciting" destinations. "You've taken your fill of the Acropolis, you've stormed the seven hills of Rome. Now...capture the city Pizarro couldn't!...visit Machu Picchu," suggested Latin American airline, Panagra Grace, in 1965.

As Fleming (1984) and Cosgrove (1994) have demonstrated, the strategies airlines employ to promote particular places often rely on highly selective projections of the world and the cartographic equivalent of artistic licence is frequently employed in an effort to communicate the worldliness and prestige of an airline's route network. As Wood (1993) similarly observed with reference to the route maps published in Delta Air Line's in-flight magazine,

> Delta's Domestic Route Map, that is, the United States and part of Canada, Mexico, and the Caribbean... [is] all but obscured beneath a thick weave of blue lines symbolising not merely Delta's *routes*, but the *embarrassing abundance* of Delta's routes. What does the map say? It says "we blanket America," that is "we will keep you so warm you will never want to go to bed with another carrier"....The point is merely to dissuade you—through the exploitation of age-old rhetorical devices (emphasis, exag-

gerations, suppression, metaphor)—from thinking of American or TWA or USAir next time you want to fly. (p. 73)

The 1970s saw a move away from the promotion of particular destinations and towards advertising campaigns that emphasized the international reach and connectivity of an airline's services. In 1974 Air France boasted connections to "161 cities in 77 countries.... See the whole world" from Paris, it suggested. Saudia, Saudi Arabia's national flag-carrier, employed an Apollo photograph of the Earth from space as a visual proxy for its route network. This emphasis on global connectivity was evident through the late 1980s. We are "Here, there, and everywhere," boasted Lufthansa in May 1987 while, in July 1988, Swissair proudly offered "over 100 landing sites on 5 continents." By the late 1990s and early 2000s, however, the effects of ongoing structural changes within the global airline industry—most notably, deregulation and the formation of code-share arrangements and global airline alliances—were increasingly evident in the advertisements that were published. Not only did advertisements for airline alliances start to appear, but individual airlines increasingly placed the logos of the alliance to which they belonged on their advertisements in an effort to emphasize their global status, prestige, and network.

In addition to marketing their services based on the range and connectivity of their international route network, airlines also sought to establish a brand that was based on positive images of customer service and their in-flight product. Such service attributes were used as the primary selling point of 44% of the advertisements that were surveyed. Basic consumer theory suggests that people purchase products according to the images they form of competing brands (see Heding, Knudtzen, & Bjerre, 2009). In the case of air travel, customers purchase a ticket with a particular carrier based not only on whether the carrier flies to the destination they wish to reach, but also based on the perceptions travelers have of the competing brands. Such perceptions are based on a range of expectations which are established through the words, pictures, and images in advertisements as well as prior experience and reputation. As all airlines essentially sell an identical product (air travel from A to B), airlines are forced to differentiate themselves from their competitors by emphasizing different attributes of their services. Unsurprisingly, the majority of the resulting advertisements revolved around notions of service, and airlines attempted to outdo one another with catchphrases which promised superlative passenger experiences, sumptuous levels of in-flight comfort, and elegant and discrete service. Panagra marketed itself as the "World's friendliest airline" (1965), while Pan Am reported that its service "Makes the going great" (1967). Elsewhere, Belgium's SABENA promised "Savoir faire in the air" (1984), while Libyan Arab Airlines promoted itself as a "24 carat airline." Very often, such promises were accompanied by photographs of female cabin crew offering fine wine and cuisine to pampered and contented-looking passengers. Far Eastern operators, in particular, were keen to emphasize the high levels of care and consideration their cabin crew bestowed upon their passengers, but European airlines, too, sought to reassure potential passengers that their every need would be attended to. "How do you get to Spain and Europe?" asked Spain's national airline in 1965, "Relaxed...with Iberia." British Airways, meanwhile, promised that "We'll take more care of you" (1980).

In addition to questions of comfort were concerns about price. From the mid-1960s onwards U.S. airlines, in particular, were keen to promote the idea that the cost of flying often compared favorably with road or rail travel, especially when the time savings were factored in. Nevertheless, air tickets remained beyond the financial reach of many and, in an effort to increase the affordability of tickets and get more people into the air, major airlines began introducing a range of new financial products, including advance purchase, apex, and tourist class fares, in an effort to stimulate demand. "Think air travel is out of bounds for your budget?" asked TWA. With our "Sky Tourist Fares" you can "save dollars." "Our Economy Tour Fares...make it easier than ever to fly away," promised Pan Am in 1967. In order to help passengers pay for their tickets, U.S. carrier TWA established a "Time-Pay Plan" that essentially enabled passengers to "fly now, pay later," and Pan Am advertised a new air travel card that allowed customers to pay in installments and thereby spread the cost of travel.

As well as arranging individual flights, airlines also promoted their own range of travel tours. Lufthansa's "pick a tour" program of the mid-1960s allowed customers to choose from 32 options, with the 30-day, round-the-world package, which included stops at Athens, Beirut, Cairo, Bombay, Delhi, Agra, Calcutta, Bangkok, Singapore, Hong Kong, Taipei, and Kyoto, reportedly being very popular and "a bargain" at $2,349. British Overseas Airways Corporation operated a similar package that would enable American tourists to sample the "best" of British history and European heritage. For $399, U.S. tourists could undertake a two-week programme incorporating eight European countries, while the all-inclusive, 15-day "Pageant of Britain" tour cost "only" $559 from New York. Through these tours, European carriers were seeking to promote inbound aeromobility from the US and encourage American tourists to visit the continent.

The third, and by far the smallest, category, which accounted for the remaining 9% of all the advertisements, concerned the utilization of new aeronautical technology, particularly the jet engine. European airlines including BOAC, British Airways, Lufthansa, and Air France used the allure of jet flight to stimulate passenger demand, and photographs and abstract images of certain jet aircraft in their fleets featured heavily in their marketing campaigns during the 1960s and 1970s. The majority of these "technological" advertisements focused on the launch of a new aircraft into revenue service. Spanish airline, Iberia, promised American passengers that their daily DC-8 service from New York to Madrid in the 1960s was operated using only the new "*extra power*" jets, while BOAC boasted that it "moves six years ahead of any other airline" on April 1, 1965 when its new, "triumphantly swift, silent, [and] serene" Super VC-10 took off for London. This emphasis on aircraft technology was designed to highlight the fact that European operators, like their U.S.-based competitors, flew modern, safe, and comfortable, aircraft.

The promotion of particular aircraft was also apparent the following decade when Air France and British Airways became the only global operators of the supersonic Concorde. Both airlines were keen to emphasize the time-saving abilities of the new technology and marketed the aircraft as something akin to time machines, with British Airways promising that "Now Super Superflight is on the way." In addition to marketing the benefits of particular aircraft, some carriers, most notably Lufthansa and Iberia, were keen to emphasize the strict maintenance regimes and employee training programs that were performed to ensure passenger safety. Spain's Iberia, for example, employed a photograph of mechanics attending to an aircraft accompanied by the reassurance that "only the plane gets more attention than you." Such direct communication ensured that the advertisement did not passively wait for its meaning to be exposed, but rather, through a combination of visual and written stimuli, grabbed the reader's attention and pushed the twin virtues of the airline's product: safety (founded on technological capability and competency) and customer service.

Irrespective of the theme into which individual advertisements were categorized, all of the airline advertisements sought to promote pleasurable consumption. This pleasurable consumption could be of the airline product itself, via exquisite levels of in-flight service and convenient worldwide connections, carefully selected and scripted foreign tourist spaces, or a combination of the two. Every advertisement sought to stimulate an emotional circuit in potential consumers so that alerting them to the prospect of obtaining a pleasurable experience created a desire to purchase the "enabling" product that would deliver it. In the case of air travel, the demand for aeromobility was derived from generating an individual or a collective desire to engage in new experiences that could not be encountered at home.

Conclusion

The enabling technology of the airplane and, in particular, the jet engine, undoubtedly facilitated the creation of new patterns of international mobility. However, these new spatialities of global aeromobility were shaped, to a large extent, by practices of place advertising and airline marketing which sought to alert passengers to new travel opportunities and promote the consumption of ever more varied and "exotic" destinations. Advertisements presented the world as the airlines wanted their consumers to see it—vast and exciting, yet also affordable and easily accessible. However, as Urry (1990) and others have convincingly argued, the manner in which individual travelers "gaze" at different spaces is highly subjective, and the particular images and perceptions that are held of different places change over time in response to new socio-economic, geopolitical, and environmental circumstances.

Airline advertisements, in common with other forms of communication media, undoubtedly reflect the dominant socio-economic and socio-cultural trends of the time in which they were produced. In the case of the examples from *National Geographic*, a clear change in advertising form and function was observed. In the mid-1960s, for example, the majority of advertisements were emphasizing the ease, safety, and utility of air travel. During the oil crises of the mid-1970s, the marketing message shifted towards the price of air tickets and the promotion of a range of flexible, "fly now, pay later" payment options that, it was hoped, would entice new passengers into the air. By the 1980s, in-flight service had become a key selling point and important product differentiator but, from the early 1990s onwards, this was displaced by advertisements that stressed seamless, worldwide air connectivity. This changing emphasis has the potential not only to inform research into the changing nature of transport and mobility, but also bears relevance for discourses on consumption and marketing.

Although based on a single source, the airline advertisements examined in this chapter illustrate the myriad ways in which post-war, international aeromobility has been promoted in print. While it is impossible to quantify the impact that exposure to these advertisements had on the travel behavior of *National Geographic* readers, the continued presence of airline advertisements over the forty-year time period suggests that airlines considered the magazine to be an important marketing and communication channel that represented an effective way to raise awareness of their brands and disseminate news of product innovations and route launches.

The advent of powered flight (and the subsequent growth of commercial air travel) during the twentieth century is often cited as a triumph of human ingenuity. Air travel facilitated globalization and enabled people, goods, capital, and information to circle the world, and the airline advertisements in *National Geographic* reflected this sense of optimism and achievement. However, recent concerns about long-term energy security and aviation's environmental sustainability have caused certain sections of global society to question our reliance on air travel and try and imagine a world in which alternative modes of travel replace aviation as the dominant mode of long-distance mobility. Given that the mobility practices of the future are likely to be very different from those of the past, printed records including, but not limited to, travel advertisements will form increasingly valuable repositories of information that will enable us to unpick the ways in which social practices of mobility and consumption have changed over time.

References

Abramson, H. S. (2010). *National Geographic: Behind America's lens on the world*. Bloomington, IN: iUniverse.com.

Adey, P. (2010). *Aerial life: Spaces, mobilities, affects*. Oxford, UK: Wiley-Blackwell.

Berelson, B. (1952). *Content analysis in communication research*. Glencoe, IL: Free Press.

Bilstein, R. E. (1984). *Flight in America: From the Wrights to the astronauts*. Baltimore, MD: Johns Hopkins University Press.

Bissell, D. (2008). Comfortable bodies: Sedentary affects. *Environment and Planning A, 40*(7), 1697–1712.

Bissell, D. (2009). Visualising everyday geographies: Practices of vision through travel-time. *Transactions of the Institute of British Geographers, 34*(1), 42–60.

Bryan, C. D. B. (1987). *The National Geographic Society: 100 years of adventure and discovery*. New York, NY: Harry N. Adams.

Budd, L. C. S. (2010). 'The view from the air': The cultural geographies of flight. In P. Vannini. (Ed), *The cultures of alternative mobilities: Routes less travelled* (pp. 71–90). Farnham, UK: Ashgate.

Budd, L. C. S. (2011). On being aeromobile: Airline passengers and the affective experiences of flight. *Journal of Transport Geography, 19*(5), 1010–1016.

Christy, J. (1987). *American aviation: An illustrated history*. Blue Ridge Summit, PA: Tab Books.

Cobham, A. J. (1928, March). Seeing the world from the air. *National Geographic, 53*(3), 348–384.

Colton, F. B. (1948, February). Our air age speeds ahead. *National Geographic*, 249–272.

Corn, J. J. (1983). *The winged gospel: America's romance with aviation*. New York, NY: Oxford University Press.

Cosgrove, D. E. (1994). Contested global visions: One-world, whole-earth, and the Apollo space photographs. *Annals of the Association of American Geographers, 84*(2), 270–294.

Crouch, T. D. (2003). *Wings: A history of aviation from kites to the space age*. New York, NY: W. W. Norton.

Cruddas, C. (2008). *100 years of advertising in British aviation*. Stroud, UK: The History Press.

Dargue, H. A. (1927, October). How Latin America looks from the air. *National Geographic, 52*(4), 450–502.

De Sieyes, J. (1918, January). Aces of the air. *National Geographic*, *33*(1), 5-9.
Edensor, T. (2003). Defamiliarizing the mundane roadscape. *Space and Culture*, *6*(2), 151-168.
Fleming, D. (1984). Cartographic strategies for airline advertising. *Geographical Review*, *74*(1), 76-93.
Gottdiener, M. (2001). *Life in the air: Surviving the new culture of air travel*. Lanham, MD: Rowman & Littlefield.
Graham, B. J. (1995). *Geography and air transport*. Chichester, UK: John Wiley.
Grosvenor, G. H. (1918, January). Germany's air program. *National Geographic*, *33*(1), 114.
Grosvenor, G. H. (1924, July). America from the air. *National Geographic*, 84-92.
Hanlon, J. P. (1996). *Global airlines: Competition in a transnational industry*. Oxford, UK: Butterworth-Heinemann.
Harrington, R. (2004). Beyond the bathing belle: Images of women in inter-war railway publicity. *The Journal of Transport History*, *25*(1), 22-45.
Heding, T., Knudtzen, C. F., & Bjerre, M. (2009). *Brand management: Research, theory and practice*. Abingdon, UK: Routledge.
Hensser, H. (1953). *Comet highway*. London, UK: John Murray.
Hewitt, J. (2000). Posters of distinction: Art, advertising and the London, Midland and Scottish railways. *Design Issues*, *16*(1), 16-35.
Jansson, D. R. (2003). American national identity and the progress of the new south in *National Geographic* magazine. *Geographical Review*, *93*(3), 350-369.
Klesius, M. (2003, December). The future of flying. *National Geographic*, *204*(6), 2-32.
London, J. (2007). *Fly now!: A colorful story of flight from hot air balloon to the 777 'Worldliner.'* Washington, DC: National Geographic.
Long, M. E. (1977, August). The challenge of air safety. *National Geographic*, *152*(2), 209-235.
Lovegrove, K. (2000). *Airline: Identity, design and culture*. London, UK: Laurence King.
Lutz, C., & Collins, J. (1991). The photograph as an intersection of gazes: The example of *National Geographic*. *Visual Anthropology Review*, *7*(1), 134-149.
Lyth, P. (2009). 'Think of her as your mother': Airline advertising and the stewardess in America, 1930-1980. *Journal of Transport History*, *30*(1), 1-21.
Millward, L. (2008). The embodied aerial subject: Gendered mobility in British inter-war air tours. *The Journal of Transport History*, *29*(1), 5-22.
Mitchell, W. (1921, March). America in the air: The future of airplane and airship, economically and as factors in national defense. *National Geographic*, 339-352.
National Geographic. (2011). National Geographic Society homepage. Retrieved from www.nationalgeographic.com
Pascoe, D. (2001). *Airspaces*. London, UK: Reaktion.
Pauly, P. J. (1979). The world and all that is in it: The National Geographic Society, 1888-1918. *American Quarterly*, *31*(4), 517-532.
Remmele, M. (2004). An invitation to fly: Poster art in the service of civilian air travel. In A. von Vegesack & J. Eisenbrand (Eds.), *Airworld: Design and architecture for air travel* (pp. 230-262). Weil am Rhein: Vitra Design.
Rothenberg, T. Y. (2007). *Presenting America's world: Strategies of innocence in* National Geographic *Magazine, 1888-1945*. Aldershot, UK: Ashgate.
Rust, D. L. (2009). *Flying across America: The airline passenger experience*. Norman, OK: University of Oklahoma Press.
Swinglehurst, E. (1982). *Cook's tours: The story of popular travel*. Dorset, UK: Blandford Press.
Szurovy, G. (2002). *The art of the airways*. St. Paul, MN: MBI.

Tuason, J. A. (1999). The ideology of empire in *National Geographic* magazine's coverage of the Philippines, 1898-1908. *Geographical Review, 89*(1), 34-53.

Tulasne, J. (1918, January). America's part in the Allies' mastery of the air. *National Geographic, 33*(1) 1-5.

Urry, J. (1990). *The tourist gaze: Leisure and travel in contemporary societies.* London, UK: Sage.

Urry, J. (2000). *Sociology beyond societies: Mobilities for the twenty-first century.* London, UK: Routledge.

Urry, J. (2007). *Mobilities.* Cambridge, UK: Polity Press.

Van Zandt, J. P. (1925, March). Looking down on Europe: The thrills and advantages of sightseeing by airplane. *National Geographic, 47*(3), 261-326.

Walsh, M. (1990). 'See this amazing America': The long-distance bus industry's use of advertising in its first quarter century. *Journal of Transport History, 11*(1), 61-89.

Watts, D. C. H. (2004). Evaluating British railway poster advertising: The London and North Eastern Railway between the wars. *Journal of Transport History, 25*(2), 23-56.

Williams, S. (1998). *Tourism geography.* London, UK: Routledge.

Wilson, J. A. (1926, October). Canada from the air. *National Geographic, 50*(4), 389-466.

Wohl, R. (2005). *The spectacle of flight: Aviation and the Western imagination, 1920-1950.* New Haven, CT: Yale University Press.

Wood, D. (1993). *The power of maps.* London, UK: Routledge.

Part II:
Mobile Selves

7

Solidarity on the Move:
Technology, Mobility, and Activism in a Hospitality Exchange Network

Jennie Germann Molz

This chapter explores the intersection between leisure mobilities, social networking technologies, and social movements through an analysis of CouchSurfing. CouchSurfing (www.CouchSurfing.org) is an online hospitality exchange network that connects global travelers who offer each other free accommodation in their homes, usually for a few nights at a time. Despite its loose network configuration, geographically dispersed membership, and usually brief meetings between members, CouchSurfing describes itself as a "global community" united around the project's mission to "create a better world, one couch at a time." The cadence of this rallying cry sounds familiar, but the logic behind this motto is far from self-evident. Indeed, it poses something of a puzzle. Under what conditions can crashing on a stranger's couch be seen as creating a better world? Furthermore, why does the CouchSurfing project appeal to a social movement discourse, as opposed to other kinds of discourses, to define itself? This vision of CouchSurfing as a global community and as a social movement raises questions about self-identity, technology, travel, mobility, and social activism that I seek to address in this chapter.

The chapter begins with a description of the background of CouchSurfing and the methods used to study the network. I then situate the project within broader debates over the social implications of new technologies to explain why CouchSurfing's appeals to solidarity and social activism are particularly puzzling. Next, I introduce research data from my fieldwork and interviews with CouchSurfing members along with close readings of the project's online Vision and Mission Statements to demonstrate how members use social movement discourse to frame their participation in the project. Drawing on social movement theory, I seek to make sense of the presumed affinity between leisure mobilities and social action that underpins the CouchSurfing ethos. The chapter concludes by suggesting that projects like CouchSurfing require us to rethink the relationship between mobility and solidarity in the context of technologically mediated social activism.

CouchSurfing: Mobile Hospitality and the Internet

CouchSurfing was originally launched in 2003 by an American web developer named Casey Fenton. Fenton got the idea for the project when he found himself with cheap plane tickets to Reykjavik but with nowhere to stay. Instead of booking a hotel, Fenton used the Internet to spam thousands of students at the University of Iceland hoping one of them would offer him a place to crash for the weekend. Within 24 hours, he had a hundred offers. This experience drew Fenton's attention to the potential of social networking and he soon launched CouchSurfing to help travelers around the world connect with available hosts. The network grew quickly, reaching 90,000 members by 2006 and more than 2.5 million members by 2011.

The online network consists primarily of members' profiles featuring autobiographical descriptions, photographs, references from previous hosts and guests that help to establish an individual's reputation within the community, and details about the member's "couch," or the kind of hospitality the member is able to offer. Because CouchSurfing is a mobile community that spans both online and offline spaces, I used a combination of face-to-face and virtual methods, including mobile participant observation in person and online and in-depth interviews with CouchSurfers. Also available online are a series of documents and member testimonials that describe how CouchSurfing works and outline the project's broader mission and vision. I conducted close readings of each of these documents. Among the themes that became quickly evident from the official documents published on the CouchSurfing website, and from members' comments posted online and shared in interviews, were a sense of resistance, solidarity, and a shared commitment to creating a better world. Clearly, members of the network see the project as more than a way to arrange free hosting. Indeed, as I have already suggested, the organization's stated objective—"creating a better world, one couch at a time"—frames travel and hospitality as globally transformative acts.

How can a loosely knit network of strangers, most of whom have never met each other, imagine itself as a grassroots social movement? How can sleeping on a stranger's couch be seen as an act of global social transformation? At best, CouchSurfing is a "complex web of hospitable ties that don't necessarily bind" (Poulston, 2011, p. 101). Yet many members make sense of their participation within the network with references to collective action, resistance, and like-minded solidarity, language borrowed explicitly from social activism. Indeed, CouchSurfing represents a fascinating paradox, made all the more intriguing against the background of a common debate: are new technologies bringing us together or tearing us apart?

Technology and the Problem of Human Togetherness

CouchSurfing entails new kinds of togetherness and imagined community based on a *combination* of online network connections and intermittent face-to-face encounters that is emblematic of wider shifts in modern social life. Indeed, among the striking features of contemporary life, especially in the wealthy societies of the global north, are the extent to which our social networks stretch across space, and the ubiquity of the information, communication, and transportation technologies we must now use to order, arrange, and mediate our sociability with distant others. According to Larsen, Urry, and Axhausen (2006), as networks of colleagues, friends, and families extend and move across geographical space, social life now involves multiple forms of co-presence established through physical travel, online interactions, and mobile communications. Thus mobile phones and the Internet, including online social networking sites, have become increasingly significant in maintaining sociality. It is becoming ever more normal to conduct our personal, family, intimate, and work lives online, on the phone, and on the move. But is this new normal of mediated and mobile communication good or bad for social cohesion?

Optimistic accounts suggest that mobile and Internet technologies facilitate, and in some cases intensify, a sense of social bonding. For example, referring to Durkheim's notion of "effervescence," a key ingredient of social solidarity, Ling (2008) argued that sharing a common mood does not require physical proximity but can be generated via mobile communication technologies. In fact, Ling suggested, mobile communication may be so effective at intensifying a group's sense of shared engrossment that it creates a kind of "bounded solidarity" to the exclusion of outside perspectives. For Chayko (2002), the Internet is similarly instrumental in establishing social bonds between people who are not physically co-present. She argued that online and mediated social interactions can create a kind of "sociomental" connection that holds virtual communities and online relationships together. Drawing on Granovetter's (1973) theory of "the strength of weak ties," Chayko pointed out that even if technologically mediated ties are themselves "weak," they demonstrate a powerful ability to "facilitate numerous opportunities for social cohesion" (Chayko, 2007, p. 378). As we will see, these concepts of effervescence, social bonding, and the strength of weak ties can help make sense of the way CouchSurfers see their online interactions and brief, co-present encounters as a form of solidarity.

At the other end of the spectrum, critics argue that even if new communication technologies are not necessarily tearing society apart, they are at the very least attenuating social ties. This is particularly evident in Bauman's (2003) concept of "liquid love," which refers to the sense that social relations are detached, loose, and fleeting. Bauman acknowledged that mobile technologies are essential to staying connected in these "liquid modern" times, but he suggested that the more we interact on the move and at a distance, the more our social relations take on the character of our digital connections: plentiful but tenuous, intense but brief, frequent but shallow. Too brief and shallow, he argued, "to condense into bonds" (Bauman, 2003, p. 62). As Bauman described it, mobile sociality revolves not just around connecting, but even more importantly around the ability to disconnect, to "unfriend," to hit "delete." Thus technologically mediated network sociality consists of relationships that are short, sweet, and easy to delete.

Furthermore, this kind of technologically mediated sociality reflects a consumerist logic that typifies contemporary social life. As Wittel (2001) observed, "connections" are currency in the network society. Bauman raised similar concerns in his discussion of the nature of human relationships in liquid modernity, arguing that the "tendency, inspired by the dominant consumerist life mode, to treat other humans as objects of consumption" signals the demise of human solidarity (Bauman, 2003, p. 75). The corrective, Bauman argued, is the "moral economy," the production of social life outside of monetary exchange. He explained that the moral economy produces an entirely different kind of sociality from the market economy—one based on solidarity, compassion, and mutual sympathy rather than frequent, fleeting, and frail connections.

Similar concerns also arise in the social movement literature, where theorists debate the limits and possibilities of new media and digital technologies as instruments of protest, resistance, organization, or democratic participation. Many theorists credit these new technologies with more flexible, decentralized, and effective configurations of social action (Melucci, 1996). They point to the way new technologies have helped mobilize concerted action in popular revolts (Rheingold, 2002), anti-war demonstrations (Bennett, 2005), anti-corporate globalization protests (Juris, 2005), and presidential elections (Owen, 2006). However, other theorists worry that new media and digital technologies threaten to splinter public attention into ever more narrow interests, thus decreasing the critical mass necessary for collective action (Brandenburg, 2006; Oates & Gibson, 2006). Due in part to its ease of accessibility and low cost of participation, the Internet has also made social movement schemas and practices available for a growing catalog of non-political, claims-making activities. The increasing use of the Internet for what Earl and Kimport (2009) referred to as "fan activism" may, in some instances, be seen as tempering the political gravitas of social activism. In the context of anti-corporate protest, online fan— or consumer-based appeals—for example, to corporate entities to change a product or save a television program slated for cancellation—seem to bear less political weight than the online coordination of radical anti-corporate globalization actions such as the World Trade Organization protests in Seattle. The fact that many of these claims-making activities take place exclusively online with no offline component also suggests that people may be taking to their computers instead of taking to the streets in a kind of "activism lite" or "slacktivism" (Morozov, 2009; see Earl, Kimport, Prieto, Rush, & Reynoso, 2010 for a useful typology of Internet activism).

From this perspective, CouchSurfing should be a prime example of "lite" activism or frail sociality. It is not making overt political or identity claims or actively reclaiming public spaces. It is not using protest techniques like petitions or boycotts, nor is it deploying any kind of traditional crowd action tactics such as marches or sit-ins. Encounters between surfers and hosts are generally short and sweet with little expectation of ongoing mutual obligation beyond the arranged stay. Furthermore, the format of the CouchSurfing website lends itself to a consumerist tendency to "shopping around" for potential hosts and guests, not unlike a dating website. As Bialski (2007) pointed out in her analysis of CouchSurfing, hospitality encounters are often framed by a logic of consumer choice that tends to commodify both people and relationships.

By all accounts, the ephemeral and loose ties among CouchSurfers and the commodifying tendencies of the website should produce a brittle and weakly integrated network. However, as we will see in the research material I introduce in the next section, this is not the picture that emerges in interviews with CouchSurfers or in the Vision Statement posted online. Instead, CouchSurfers refer quite explicitly to concepts that seem less market-oriented and more indicative of Bauman's "moral economy": emotional intensity, shared compassion, generosity, mutual support, and a sense of solidarity. In the analysis that follows, I draw on various concepts from social movement theory in order to make sense of this apparent paradox.

One Couch at a Time

In this section, I discuss the way the CouchSurfing website and CouchSurfers themselves draw on social movement rhetoric to frame their participation in CouchSurfing as a form of resistance (against a mass-mediated culture of fear or against the corporatization of social life) and as a form of positive action toward creating a new order of social life.

Resistance

References to CouchSurfing as a form of resistance first emerged in interviews with female CouchSurfers in the United States who told me that they saw their membership in CouchSurfing—and in particular their willingness to trust strangers—as a resistance to a sense of fear and "stranger danger" circulated in the mass media. In response, they imagine a world characterized by trust in strangers, a vision they enact by surfing with and hosting strangers. Dylan, a graduate student in her twenties from New York, explained that:

> CouchSurfing...does reinstate, for me, a faith in a sort of humanity. Not a faith in humanity, that's way too dramatic. But just the idea that you can trust strangers....That you can be friends with them. That you don't have to put up these walls. People are people everywhere, and we have such this culture of fear. I feel like this is a kind of site that tries to sneak in through the back of that culture of fear. It's not buying it. [It's] creating pathways out of that. Or around it.

Lara, a yoga instructor in her early thirties and mother of two small children, blames this perception of fear and danger directly on the media. She tells me that some of her friends think she is crazy for CouchSurfing, but she rationalizes their concern by explaining:

> people are just afraid...I think too many people are so media clouded that...all they think is "Fear and terror and oh my god what could *happen*?" Instead of, "Oh, what *could* happen?" Like "Hmmm, it might be fun!" instead...I think the media definitely tries to scare us and keep us from doing a lot of things.

For Dylan, Lara, and many other respondents, CouchSurfing is a way of speaking back to a mass-mediated culture of fear by seeing strangers and unfamiliar situations as potentially fun and fulfilling rather than potentially dangerous.

A different connotation of resistance also emerged in other interviews, this time as "resisting consumer culture." For a few respondents, CouchSurfing's status as a non-profit organization and its ability to facilitate a non-commercial form of exchange between strangers represented a resistance to what they saw as a general corporatization of social life. This was particularly important to Nico, an artist I met in Italy. I met Nico and his girlfriend, Marie-Eve, both in their early thirties, in the small artists' colony where they live in northern Italy.[1] We talked for several hours about the Internet, the evils of the market economy, and the revolutionary potential of networks like CouchSurfing:

> Nico: The beauty of these Internet sites and programs, in my opinion, is that they all serve a huge revolution. The revolution to no longer use money. This is the real aspect of the revolution that is embedded in these social networks.

Marie-Eve: Yes, you escape the capitalist consumer system.

Nico: Because the creators of the Internet, the creators of "open source," know very well that our common enemy is the Federal Reserve....These websites make something more beautiful possible: to travel the world without money.

For Nico, CouchSurfing represents the revolutionary potential of the Internet precisely because it facilitates an economy of sharing, mutual help, and generosity that operates outside of the money economy. According to Nico, using the Internet to facilitate non-commercial exchanges of any kind—such as free hospitality and ride-shares—is what the Internet was meant to be used for. He tells me that websites like CouchSurfing will eventually "demonstrate that we're all one family. One big family."

Indeed, hospitality exchange sites like CouchSurfing hark back to the early principles of non-commercial, democratic, peer-to-peer communication, and community. In describing its non-profit status, the CouchSurfing website reiterates: "The goal of CouchSurfing has never been about money.... It's all about helping to reach our vision of a better world" (http://www.CouchSurfing.org/ about.html). By rejecting profit models and commercial exchange, CouchSurfing reasserts the "true" intentions of the Internet: to create a global village of strangers meeting strangers. This non-commercial ethos echoes some of the rhetoric surrounding the Internet and the virtual communities that were forming on bulletin boards and multi-user domains in the early 1990s. Utopian thinkers at the time suggested that virtual communities could be more democratic and inclusive. As Juris (2005, p. 191) explained in his study of digital media use in anti-corporate social movements, many activists see the technological architecture of the Internet, especially the open-source development process, as a "broader model for creating alternative forms of social, political, and economic organization." In other words, the Internet is both an instrument for organizing social activism and a model of democratic, non-corporate, post-capitalist forms of collaboration, community, and social organization (Juris, 2005, p. 192).

In this sense, CouchSurfing appears to fulfill the original utopian promise of the Internet to unite strangers across geographical and cultural divides and to form a global community. In the case of CouchSurfing, however, this community may connect online, but it coalesces offline. Nico, along with several of the other CouchSurfers I encountered during my fieldwork, saw CouchSurfing as a tool for living outside of the consumerist grid. Along with CouchSurfing, they participate in non-commercial exchanges such as bartering and hitchhiking to coordinate their material and social lives. Beth, a CouchSurfer in her late thirties, makes the point in our interview that such freely exchanged resources also help to legitimize these otherwise anomalous choices:

> It seems to me like most people I know who are travelers...are kind of looked down on by family members. Or people question their choices, like, "Why don't you have a steady job? Why don't you have a home?"...It seems to me that something like CouchSurfing is creating a support network for people like that, which I think is really great. And also kind of legitimizing that way of being, whether it's for a few days, a few weeks, or your whole life.

Through websites like CouchSurfing, the Internet is mobilized in support of alternative lifestyles that are on the move and outside the corporate framework. These CouchSurfers see their participation in the network as a form of collective resistance to the fear and consumerism that dominates Western society. But the CouchSurfing project, to the extent that it is identified by members as a form of anti-corporate protest at all, is framed as a far more moderate response to the market economy than, say, the World Trade Organization and G8 protests in Seattle, Genoa, and other cities (Juris, 2005). It is perhaps more akin to recycling than rioting.

In the absence of overtly political claims, CouchSurfing represents what Dalli and Corciolani (2008) referred to as "moderate" social activism. In their analysis of collective forms of consumer resistance in online communities, Dalli and Corciolani compared CouchSurfing to other websites that facilitate non-monetary exchanges among members, such as BookCrossing, an alternative, online book exchange. They argued that CouchSurfing and similar websites are not aimed at market subversion, but rather at using "the market in order to meet new needs: solidarity, democracy, consumers' emancipation" (2008, p. 757). Despite what some critics regard as a consumerist veneer, CouchSurfing's appeal to non-commercial relations generates a sense of belonging and solidarity among some of its members. In practice, though, leisure travel and hospitality are much less about explicit conflict than they are about solidarity at the margins of consumer culture and corporate globalization. Staying with and hosting strangers is, for these CouchSurfers, a way of supporting and normalizing alternative lifestyles as a means to creating a better world.

Creating a better world

The notion of "creating a better world" is a common refrain, not just in interviews, but on members' profile pages, in online testimonials, and in the Vision Statement posted on the CouchSurfing website (http://www.CouchSurfing.org/about.html/vision). In fact, the Vision Statement outlines in detail the conceptual leap from "crashing on someone's couch" to "creating a better world" with the following flow chart: explore → connect → appreciate diversity → global community.

Taking each of these four values—explore, connect, appreciate diversity, global community—in turn, the Vision Statement explains that connections between strangers can establish a foundation of empathy based on what individuals have *in common* so that they can learn about and understand their *differences* from a perspective of respect and appreciation. According to the statement, to "explore" is to "open ourselves up to the unfamiliar with a sense of wonder." Next, CouchSurfers "connect," not just with people who are similar, but with people who are very different. CouchSurfing's objective is to facilitate meaningful and inspiring connections by offering a "fun, safe, and easy way to meet people from all over the world" and to "compare our experiences and find comfort in our similarities." Once CouchSurfers have explored and connected, they then "appreciate diversity." As the Vision Statement explains, "we can begin to view our differences with understanding and compassion....We can develop empathy for our differences and even learn to better accept ourselves."

The Vision Statement defines what a better world looks like: "Imagine living in a world where there are no sides and all kinds of philosophies and ideologies can coexist....Imagine living in a world where we more fully appreciate unfamiliar perspectives and other people are more appreciative of ours." At this point, CouchSurfers help to realize a global community, which the Vision Statement describes as:

> a world where everyone is inspired to help and care for each other, regardless of differences in culture or ideologies....Wouldn't our feeling of kinship across cultures also make it impossible to accept aggression—no matter whom it's directed towards?... Can you imagine a world filled with people who have a feeling of connectedness, committed to helping each other? Can you envision yourself as part of a critical mass of people extending hospitality, sharing knowledge, spreading tolerance and supporting each other as we strive to meet our collective needs?....We believe that relationships we build across continents and cultures can create a global community that values diversity and seeks understanding in times of disagreement. *Will you join us as we pursue this vision?*

In the ideal world envisioned in these documents, empathy, tolerance, respect, mutual help and care, and a sense of kinship are the building blocks of a new sociality. This is the cornerstone of a new social order and a new global community—in other words, a better world—where, according to the Vision Statement, "everyone is inspired to help and care for each other, regardless of differences in cultures or ideologies."

To return to my earlier question: How is it that solidarity and even a new sense of kinship emerge out of such a loosely integrated network of individuals who interact only temporarily online and meet up only briefly and rarely, if at all? How do we make sense of social relations among strangers that are, as Bauman (2003, p. xi) put it, "light and loose," plentiful, tenuous, and brief, and *at the same time* enacted within a discourse of generosity, empathy, emotional intensity, collective needs, and mutual compassion, support, and help?

One way of understanding this is to consider the link between the individual and the collective. In the Vision Statement, as well as in my interviews, CouchSurfers suggest that the project is changing the world by changing individuals. It is clear from the Vision Statement that connections with strangers are opportunities for CouchSurfers to learn about themselves and others while cultivating personal qualities of empathy, open-mindedness, respect, and appreciation. In her analysis of CouchSurfers' motivations to travel, Bialski (2007) also identified this link between personal growth and "global utopia." She noted a very common belief among CouchSurfers that the world can be transformed through individual experiences of personal growth. Bialski saw the self-reflexivity underpinning this personal growth (i.e., learning about one's self) as a "commitment which is part of a Utopian ideal to better society (or even, the world) through bettering oneself" (2007, p. 47). But how does the highly reflexive practice of personal growth translate into collective action?

The Vision Statement crafts a meticulous logic that frames personal development and self-knowledge as individual pieces of a larger constellation. For example, it suggests that a new global social order relies on a "critical mass of people" working to meet "collective needs." Member testimonials posted throughout the Vision Statement illustrate this logic of collective action by emphasizing the larger effects of individual-level connections or referring to themselves as individual pieces in a big puzzle:

> If the human race is to survive our political blunders, ignorance, and obstinacy it will be because groups such as CouchSurfing have created a way to connect on an individual level. The foreigner across the ocean is retitled a friend, the other is the same, and the experience of sleeping in someone's house and seeing through their eyes fundamentally changes the collected atoms that make up who we are. (Member Testimonial: Argentina)

> Since I discovered CouchSurfing, I have learned that the world has millions of different points of view ready to be discovered thanks to the human value of hospitality. (Member Testimonial: Spain)

Members are told that "if enough of us" participate and if we have "enough of these experiences," then we "may begin to see a world where people feel a greater sense of connection with each other, in spite of differences." In this way, a very loosely woven network of *atomized* individuals is imagined as a collective *whole* with transformative potential.

The literature on New Social Movements offers some clues to understanding this extension of individual reflexivity to collective action. New Social Movements emerged in the 1960s, especially in Western societies, out of student movements: new wave feminism, the peace movement, the environmental movement, and so on (Melucci, 1996). Crossley (2002) explained that what these movements have in common is precisely the fact that they do not address overtly political projects and are often concerned with supporting alternative lifestyles as a form of resistance to objectionable, dominant cultural modes. According to Crossley (2002, p. 152), New Social Movement theory focuses on change at the cultural, symbolic, and sub-political domains, taking seriously the feminist slogan that "the personal is political." Melucci (1996) argued that in these movements:

> solidarity is cultural in character and is located in the terrain of symbolic production in the everyday life. To an increasing degree, problems of individual identity and collective action become meshed together: the solidarity of the group is inseparable from the personal quest and from the everyday affective and communicative needs of the participants in the network. (p. 115)

Taking this logic further, CouchSurfing might also be considered what Crossley (2003) called an "even newer social movement." Crossley introduced this new category to describe the new anti-corporate movement he saw taking shape not only in the form of highly visible demonstrations against the World Trade Organization or the International Monetary Fund, but also among thousands of smaller organizations that organize localized protests and alternative projects. By offering a non-commercial alternative to corporate forms of hospitality, CouchSurfing might certainly count among this number. In this sense, social movements do not just rely on collective action, but also self-change to pursue their ideals. This framework implies that hospitality encounters can be imbued with political significance, and sheds light on the reasoning that links personal development to the creation of a new global social order (see also Bialski, 2007).

Emotions are also central to the process of translating personal experiences into collective action, lending a sense of coherence to an otherwise dispersed and disparate membership. For example, the opening comments of the Vision Statement introduce the document as something that "should resonate with all members of the organization and help them feel proud and excited to be part of something bigger than themselves." In this sense, the Vision Statement calls on certain emotions (resonance, pride, and excitement) to structure a sense of collectivity within the network. In the case of CouchSurfing, emotions are not just a structural device. They constitute the cause itself. Social movements are often characterized by such appeals to emotions, which serve to embed activists in "social networks that extend across space and over time" (Bosco, 2006, p. 343). The mission of CouchSurfing is to encourage individuals to *feel* a particular way (Bialski, 2007). It aims to produce specific emotions, such as empathy and tolerance, or to overcome other emotions, such as fear and inhibition. Two additional member testimonials posted on the website's Vision Statement highlight the way CouchSurfing makes them *feel*:

> CouchSurfing is a piece of heaven. It makes you feel your sense of humanity deeply. (Member Testimonial: Turkey)

> You know that feeling when you truly connect with another human soul—when you find yourself talking to a stranger who feels like a friend—a new friend who totally inspires and amazes you—and you just feel energized and so alive?! Well, CouchSurfing connects you with those people, all around the world or even in your own home. (Member Testimonial: New Zealand)

The new global social order is thus constituted by the way strangers feel about one another. Feelings of faith, trust, curiosity, empathy, compassion, friendship, and, above all, a feeling of connectedness, characterize the ideal world that CouchSurfing asks its members to imagine. There is certainly evidence of what Durkheim called "effervescence" in the Vision Statement and the member testimonials excerpted above. These shared emotions—of resonance, excitement, feeling a deep "sense of humanity," or feeling "energized and so alive"—along with the emotional intensity of the CouchSurfing encounters themselves (Bialski, 2007) may generate a sense of solidarity across this highly dispersed and loosely integrated network (Ling, 2008). In this sense, solidarity and emotions are ends in themselves as much as they are a foundation for further social action.

Moderate social activism, New Social Movement theory, and the role of emotion in social movements help us to decipher, to some extent, the conditions under which crashing on a stranger's couch can be seen as creating a better world. For one thing, these frameworks outline some of the mechanisms by which individual expressions of moderate resistance might translate into collective action. For another, they suggest that it may be the social movement itself that generates a sense of solidarity, rather than the other way around. In other words, CouchSurfing draws on social movement discourse to hold an otherwise loose and dispersed network of strangers together.

Solidarity, in particular, is thus both a condition for and an effect of "creating a better world." But this solidarity is not like traditional forms of solidarity, which are commonly assumed to emerge over time within tightly knit, place-based communities. Instead, as I will describe in the next section, CouchSurfing revolves around a kind of "mobile solidarity" generated through geographically dispersed, asynchronous, and networked online and offline interactions between strangers. The concept of mobile solidarity also allows us to reconsider the complex role of *movement* in social movements (Sheller, 2001) and to better understand the particular relationship between social movements and leisure mobilities on which the CouchSurfing ethos is founded.

Mobilizing Solidarity

In thinking about the role of movement in CouchSurfing, it is useful to consider the historical legacy of hospitality exchange networks. Despite the fact that several of the members I interviewed framed CouchSurfing as a resistance to commercialism, CouchSurfing is less the offspring of the anti-corporate, globalization protest movement than it is a descendant of the peace movement. It follows in the footsteps of Servas International, a hospitality network founded in 1949 by Bob Lutweiler to promote tolerance and world peace through person-to-person interactions. By the 1990s, several hospitality exchange organizations appeared online, including Hospitality Club, Hospitality Exchange, and, of course, CouchSurfing. Like Servas, these networks were guided largely by the belief that world travel, interpersonal exchanges between people from diverse cultures, and the generosity expressed through free hospitality could spread tolerance, friendship, and world peace at a grassroots level. Deriving from this movement, CouchSurfing clearly situates travel at the center of its project. The forms of resistance, global transformation, and solidarity described earlier are predicated, quite literally, on the corporeal and technological mobility of the network's members. What we need, then, is a way of thinking about CouchSurfing that accounts for the complex interplay of leisure mobilities and online networking within a social movement schema. To paraphrase Sheller (2001), we need to examine the role of *movement* in social movements. To this end, I argue that CouchSurfing constitutes a kind of "mobile solidarity" that requires new ways of conceptualizing both solidarity and social movements.

Sheller (2001, p. 2) noted that the study of social movements must inevitably "consider how social movements engage in both the literal motion of bodies and things through space and with the 'virtual mobilities' afforded by new information and communication technologies." Because social movements are literally and virtually in motion in these variable and intersecting ways, Sheller argued that social movements are better understood in terms of "social flows" rather than through structural metaphors such as networks and social ties. In contrast to structural metaphors, which beg the question of a "prime mover" that instigates the movement in the first place, Sheller argued that our question should be: "how are social movements constituted by the relational settings in which they occur, and how, in turn, do these relations (re)constitute a political context?" (2001, p. 4). In other words, she argued, social movements are not necessarily instigated by "governing" actors; instead possibilities for mobility arise out of the relational dynamics of a system in which contexts, actors, and mobilization are contingently produced. Sheller also suggested that a social flow approach to social movements enables us to see "social actors and contexts as *effects* of mobilization rather than simply as conditions" (2001, p. 3). By shifting the question from catalysts and outcomes to contexts and relations, Sheller framed mobility as both a condition and an effect of social movements. Following Sheller's approach, I want to conclude this chapter by speculating on a few of the mechanisms by which CouchSurfing is constituted as a form of mobile solidarity.

First, CouchSurfing is hybrid. It involves both the corporeal mobility of bodies, materials, and technologies as well as the virtual mobilities of information and communication. Online social networking makes it possible for individual members to find and communicate with each other, while the website and the open discussion forums within it become virtual spheres for distributing and debating the network's shared values and goals, allowing a collective identity to emerge. Even though most of the members of CouchSurfing will never actually meet each other in person, the website provides a central meeting place and a sense of coherence across the network, thus creating its own context. Technologically mediated forms of communication are usually associated with living social life "at a distance." In the case of CouchSurfing, they underpin a more nuanced intersection between distance and proximity. Social interactions at a distance are intended to culminate in local interactions between strangers. Crucial to a sense of mobile solidarity, therefore, is the way CouchSurfing translates online connections and online participation into offline relationships and offline action. According to media artist and theorist, xtine, CouchSurfing is thus not a moderate form of social action, but rather a radical medium for "local acts of transgression" (xtine, 2008, para. 15). She noted that members are urged to "participate both on and offline to create cultural exchange and to dissent against corporate cultural governance," such that CouchSurfing represents a new use of interactive network technology to "promote alternative solutions, locally" (para. 22). According to xtine, CouchSurfers are thus "empowered online as content generators and offline as activists living by the site's creeds" (para. 22). In this case, mobile solidarity is hybrid, flowing between online participation and offline action.

Second, CouchSurfing is not centrally coordinated. CouchSurfing involves dispersed acts of hospitality, not large displays of collective action. Each individual hosting encounter—of which there are thousands across the globe every week—is an articulation of the project's mission that is only ever realized at a local level. Each meeting between CouchSurfing members contributes to the "critical mass" that the organization deems necessary to "create a better world." In turn, these encounters help people "feel a greater connection" and sense of solidarity. At the same time, these encounters are notoriously brief, usually lasting only a day or two. In some cases, these meetings evolve into ongoing friendships, but even if they do not, these ephemeral encounters help constitute a broader sense of solidarity.

To a certain extent, the decentralized nature of CouchSurfing as a social movement aligns with Melucci's (1996) observations that new information and communication technologies are transforming the structure of social movements and the conditions of mobilization, with traditional images of the "politically organized actor" giving way to a more diffuse leadership and membership. New technologies lend a more amorphous and web-like character to social movements. Melucci (1996) explained that contemporary movements are thus "segmented, reticular, and multi-faceted," with "individual cells operat[ing] on their own entirely independently of the rest of the movement, although they maintain links to it through the circulation of information and persons" (p. 115). Melucci's descriptions gestured toward the more fluid nature of social movements, though his metaphors of networks and cells remains focused on individual ties.

Instead, Sheller's metaphor of waves and particles more aptly captured the sense of social movements as flows. As Sheller (2001, p. 11) proposed, social movements might be understood through the "classic quandary of light and other quantum entities behaving both as particles and as waves." She noted that if we only focus on the particles—individual actors, or the individual social ties between actors—we miss vital elements of the social flow that makes up social movements. In the case of CouchSurfing, just because the distinct interactions between members of the network appear to be brief and weak does not mean that the organization as a whole is frail. It may actually be quite flexible and resilient.

Finally, CouchSurfing reflects the ways in which mobility is both an effect of and a condition for solidarity. Sheller's (2001, p. 8) observation that "rather than a particular end-state, movement itself may be a movement's *raison d'être*" seems especially salient in the case of CouchSurfing, where movement quite literally constitutes the movement. The logic that links crashing on strangers' couches to a new global social order is structured around physical mobility. By traveling, CouchSurfers expand their perspectives, cultivate their sense of compassion, and create a better world. But, as Sheller might argue, travel should not be mistaken for a "prime mover" here. Travel helps to constitute the CouchSurfing community, but the CouchSurfing community also enables travel. The network acts as a support system for travel, encouraging more and more members to carry out the project's vision. The end-state, as described on the website, is not stasis, but even more movement. The goal is to create more ways for people to travel, explore, and connect. Solidarity and social activism are both a context for and an effect of this mobility. The more CouchSurfers travel and connect, the more a sense of solidarity develops across the network. At the same time, an already existing sense of solidarity or "like-mindedness" is what reassures some CouchSurfers and motivates them to travel. Some might even argue that CouchSurfing's vision of cultivating tolerance and compassion is flawed from the outset because the only people who would join a networking site like this are already open-minded and empathetic.

The interplay between mobility and solidarity may also help to explain why CouchSurfers draw on a social movement discourse, as opposed to other discourses, to make sense of their participation in the project. To a certain extent, the members of CouchSurfing make sense of their connections with each other as a form of social activism. "Social activism" becomes a format for individuals to connect to each other on the move; it is the social movement discourse *itself* that holds together an otherwise loose and dispersed network of strangers. "Creating a better world, one couch at a time" is both an organizational mission and a rhetorical context within which online and face-to-face interactions between strangers make sense. Earl and Kimport's (2009) research on fan activism is instructive here. They observed that groups of like-minded fans or supporters will often appeal to social movement vocabulary and employ social movement tactics for non-political cultural or lifestyle claims.

For example, fans will protest the cancellation of a favorite television program by using tactics such as petitions or boycotts. Even without advancing strictly political claims, the language and tactics of social activism can give structure to a dispersed community. Citing Anderson (1983), Earl and Kimport (2009, p. 236) explained that even when "these tactical implementations focus on claims far from traditional social movement concerns, [they] often give voice to classes of individuals—usually fans—and their 'imagined communities.'" The rhetoric of social activism itself can frame participants as a community of like-minded individuals and give shape to a collective identity (Earl & Kimport, 2009, p. 237). Mobile solidarity is thus a context for and an effect of practicing hospitality as a social movement.

Conclusion

As I have described throughout this chapter, many of the CouchSurfers I spoke with described their acts of hosting and surfing as forms of resistance to the distancing forces in modern social life: a mass-mediated culture of fear that vilifies strangers, the commercialization of the Internet, or the corporatization of social life. In turn, they associated their participation in CouchSurfing with a desire to "create a better world." These claims, that hosting a stranger or crashing on their couch can literally change the world, reframe hospitality not just as a leisure pursuit, but as a form of social action. This is not to say that all CouchSurfing encounters operate according to this ethos, or that all members of the network subscribe to the project's broader mission—many really are just looking for a free place to crash. For the most part, however, the discourses I have described throughout this chapter work in powerful ways to structure the way members participate in the network and make sense of their encounters with strangers.

Throughout this chapter, I have engaged several concepts drawn from mobile communication studies, social networking analysis, and social movement theory in order to understand the logic by which crashing on someone's couch can be seen as creating a better world. First, I argued that CouchSurfers mobilize weak ties (Granovetter, 1973) in support of alternative lifestyles, and that such lifestyle claims can be understood as expressions of political resistance against "stranger danger" or the market economy. Drawing on New Social Movement theories, I then explained how CouchSurfing translates self-reflexivity, personal growth, and individual feelings into a sense of solidarity and collective action, paradoxical though that may seem. I also suggested that CouchSurfing appeals to social movement discourse, including shared emotions akin to Durkheim's "effervescence" (Ling, 2008), in order to extend a sense of coherence across this dispersed and technologically mediated community of strangers.

Finally, drawing on Sheller's (2001) notion of "social flows," I explored the assumed affinity between leisure mobility and social movements embedded in the CouchSurfing Vision Statement, showing how physical movement, virtual mobilities, and solidarity are co-constituted in the discursive, digital, and material practices of CouchSurfing. From this perspective, we can think of CouchSurfing as a form of mobile solidarity constructed around flows rather than networks. As I have suggested, solidarity seems an unlikely description for CouchSurfing, with its mobile membership and intermittent encounters. Unlike conventional solidarity, mobile solidarity is based on brief, dispersed, and decentralized acts of online and offline interactions between members. The Internet and sophisticated networking technologies are central to these hospitality encounters, enabling new contexts for social action, new avenues toward a "moral economy" of shared empathy and generosity, and new ways of negotiating solidarity among mobile strangers. Mobile solidarity does not necessarily answer the question of whether new technologies are cohering or fragmenting society—if anything, it poses new forms of inclusion and exclusion—but it does allow us to imagine new possibilities for togetherness on the move and for social action at a distance that give solidarity renewed significance in a mobile world.

Acknowledgements

The author wishes to thank Jim Conley, members of the Department of Sociology and Anthropology at Holy Cross, and fellow CouchSurfing researchers for their valuable comments on earlier drafts of this chapter.

Notes

1. These respondents requested that I use their real names in my analysis, which I have done. All other names have been changed.

References

Bauman, Z. (2003). *Liquid love: On the frailty of human bonds*. Cambridge, UK: Polity Press.

Bennett, W. L. (2005). Social movements beyond borders: Understanding two eras of transnational activism. In D. Della Porta & S. Tarrow (Eds.), *Transnational protest and global activism* (pp. 203–226). Lanham, MD: Rowman & Littlefield.

Bialski, P. (2007). *Intimate tourism: Friendships in a state of mobility—The case of the online hospitality network*. (MA Thesis). University of Warsaw, Warsaw, Poland. Retrieved from: http://intimatetourism.files.wordpress.com/2007/07/paulabialski-thesisma-intimatetourism.pdf

Bosco, F. (2006). The Madres de Plaza de Mayo and three decades of human rights' activism: Embeddedness, emotions, and social movements. *Annals of the Association of American Geographers, 96*(2), 342–365.

Brandenburg, H. (2006). Pathologies of the virtual public sphere. In S. Oates, D. Owen, & R. Gibson (Eds.), *The Internet and politics: Citizens, voters and activists* (pp. 207–222). New York, NY: Routledge.

Chayko, M. (2002). *Connecting: How we form social bonds and communities in the Internet age*. Albany, NY: SUNY Press.

Chayko, M. (2007). The portable community: Envisioning and examining mobile social connectedness. *International Journal of Web Based Communities, 3*(4), 373–385.

Crossley, N. (2002). *Making sense of social movements*. Buckingham, UK: Open University Press.

Crossley, N. (2003). Even newer social movements? Anti-corporate protests, capitalist crises and the remoralization of society. *Organization, 10*(2), 287-305.

Dalli, D., & Corciolani, M. (2008). Collective forms of resistance: The transformative power of moderate communities: Evidence from the BookCrossing case. *International Journal of Market Research, 50*(6), 757-776.

Earl, J., & Kimport, K. (2009). Movement societies and digital protest: Fan activism and other nonpolitical protest online. *Sociological Theory, 27*(3), 220-243.

Earl, J., Kimport, K., Prieto, G., Rush, C., & Reynoso, K. (2010). Changing the world one webpage at a time: Conceptualizing and explaining 'Internet activism.' *Mobilization, 15*(4), 425-446.

Granovetter, M. (1973). The strength of weak ties. *American Journal of Sociology, 78*(6), 1360-1380.

Juris, J. (2005). The new digital media and activist networking within anti-corporate globalization movements. *The Annals of the American Academy of Political and Social Science, 597*(1), 189-208.

Larsen, J., Urry, J., & Axhausen, K. W. (2006). *Mobilities, networks, geographies*. Aldershot, UK: Ashgate.

Ling, R. (2008). *New tech, new ties: How mobile communication is reshaping social cohesion*. Cambridge, MA: MIT Press.

Melucci, A. (1996). *Challenging codes: Collective action in the information age*. Cambridge, UK: Cambridge University Press.

Morozov, E. (2009, May 19). The brave new world of slacktivism. *Foreign Policy*. Retrieved from http://neteffect.foreignpolicy.com/posts/2009/05/19/the_brave_new_world_of_slacktivism

Oates, S., & Gibson, R. (2006). The Internet, civil society and democracy: A comparative perspective. In S. Oates, D. Owen, & R. Gibson (Eds.), *The Internet and politics: Citizens, voters and activists* (pp. 1-19). New York, NY: Routledge.

Owen, D. (2006). The Internet and youth civic engagement in the United States. In S. Oates, D. Owen, & R. Gibson (Eds.), *The Internet and politics: Citizens, voters and activists* (pp. 20-38). New York, NY: Routledge.

Poulston, J. (2011). Conference report: CHME. *Hospitality & Society, 1*(1), 99-105.

Rheingold, H. (2002). *Smart mobs: The next social revolution*. New York, NY: Basic Books.

Sheller, M. (2001). *The mechanisms of mobility and liquidity: Re-thinking the movement in social movements*. Lancaster, UK: Department of Sociology, Lancaster University. Retrieved from: http://www.comp.lancs.ac.uk/sociology/papers/Sheller-Mechanisms-of-Mobility-and-Liquidity.pdf and http://www.lancs.ac.uk/fass/sociology/papers/sheller-mechanisms-of-mobility-and-liquidity.pdf

Wittel, A. (2001). Toward a network sociality. *Theory, Culture & Society, 18*(6), 51-76.

xtine. (2008, April). CouchSurfing, Delocator, and Fallen Fruit: Websites respond to a crisis of democracy. *M/C Journal: A Journal of Media and Culture, 11*(1). Retrieved from: http://journal.media-culture.org.au/index.php/mcjournal/article/viewArticle/24

technologies and mobility infrastructures, in part through holding required motility skills and resources.

So we have extensions, but what is the self doing with them? Across our lives we have needs, and from an everyday life perspective, they can be seen as changing across the life course. In earlier times, with technologies that did not enable the extended reaches possible today, needs had to be addressed within smaller, less robust extensions, closer to humans' original naked capacities. Hall saw life as a dance where people are tied together, or isolated from one another, by invisible rhythms. For him, human extensions can be viewed as externalized manifestations of human drives and needs (Hall, 1983, p. 130). Kellerman (2006) included "personal needs for mobility" in his basic model of mobility. Relating needs to settlement patterns, Jacobs (1961/1992, p. 230) argued that a lack of a wide range of concentrated diversity can put the self into a car, creating a car and road extension of the self, for almost all of one's needs, instead of a self walking on a sidewalk extension, to address needs locally. One can go further still by considering how competing needs of the self, and others with whom arrangements are built, are addressed and result in actualized connections.

Towards Tracing Motility of the Networked Self: Mobility Action Chains

Through an introduction to the networked self and the various extensions that comprise it, it is possible to turn one's attention now to tracing potential mobilities in the form of chains of connections that bring the self together with needs. Burgess (1925/1967, pp. 58–59) described mobility as the "pulse" of the community and demonstrated how it is indicative of all the changes taking place in the community. Canzler, Kaufmann, and Kesselring (2008, p. 2) provided multiple meanings to the term "tracing mobilities," with one being their intention to trace out how mobilities are inscribed into different spheres of modern life. Importantly, both Burgess and Canzler and colleagues saw movements through the use of mobility technologies and mobility infrastructures across geographical space as having a social component, and they attempted to describe and measure this movement in some way. In an attempt to refine these notions of pulses and tracings, one can consider mobility action chains utilizing Hall's (1976/1989) "action chains" as a foundation. If there is potential for a mobility action chain to succeed, then a form of motility exists. If a mobility action chain is not possible, then motility is not possible through a specific chain configuration. Hall doubted whether humans could do anything of a social nature that did not involve action chains. At the same time Hall recognized the varying scales and temporality of these chains. In considering physical space, Hall suggested that action chains can be a useful tool for architects and planners, noting that research on how spaces are used "reveals that failure to get detailed data on the action chains and the situational frames in which they occur can result in breaking the chain" (Hall, 1976/1989, p. 141). Here we are reminded of Bijker and Law's (1992) seamless (what can be considered as a successful mobility action chain), and at times not so seamless (what can be considered as an unsuccessful or reconfigured mobility action chain), web of connections. Hall's action chains are located within what he referred to as "situational frames," which include movement, spatiality, temporality, sociality, and materiality, among others. Importantly, Hall suggested that frames represent the materials and the context in which action occurs.

At one end of a mobility action chain is the central portion of the networked self that can change, based in part on changing skills, access, appropriation and coping strategies. There may also be "others" involved in enhancing skills and access that would not otherwise be possible. At the other end of a mobility action chain are the things that are sought out in an attempt to fulfill needs. Contributing to bringing a particular potential mobility action chain to life are mobility technologies and mobility infrastructures and built environments, acting as a stage where actions can take place. The mobility action chain may involve one moving physically, or alternatively some proxy in a physical and/or virtual sense, or some combination thereof. Focusing in on everyday life mobility, Jensen (2009a) saw a social environment within what Shane (2005) described as armatures (resembling what we can consider as settings of mobility technologies and mobility infrastructures) as a place of meaningful interaction. Here, presumably, one who is immobile, or has restricted mobility, would not be able to actualize mobility action chains that enable participating in social environments within armatures (Shane 2005), nor cultures of movement, to the same degree as those who are fully mobile. Yet, through arrangements with others, perhaps there are ways to participate. Mobility action chains can fall within a wide spectrum of possibilities, from being clean, crisp, and easily identified, through to being highly complex, with frames within frames and chains within chains. Mobility action chains can also encompass competing and conflicting needs, not only of an isolated individual self, but also of others with whom arrangements are built. As a part of possible coping strategies, mobility action chains may be reconfigured. Here we can consider Malabou's (2008, p. 10) notion of the plasticity of actions within networks, where one needs to be connected, modulating one's efficacy and retaining suppleness, or risking rejection.

With our theoretical frame in position, it can be utilized as a lens in considering glimpses of motility of the networked self. Attention will be drawn to the self, starting as a newborn, through youth, midlife, late-life, and finally end of life. With our limited time and space, vignettes of everyday life mobility and motility at these various stages of the life course are presented. The vignettes bring together work by other scholars and commentary on how the theoretical frame presented herein can be used as a lens towards considering motility of the networked self across the life course.

Glimpses Across the Life Course

Newborn

Many have heard stories of parents rushing to the hospital and the odd story of births taking place en route to a hospital. For others it is a long wait by parents, family, and friends in a hospital or some other setting awaiting birth. Our newborn enters the world naked, without any technological extensions. As an infant, potential independent mobility is quite limited in comparison to those around us, such as parents and health care workers. If all goes well, a newborn is taken home to settle into life. At this stage needs are brought to a newborn by others. Through mobile other arrangements, which rely on the motility of others, mobility action chains bring the essentials of life to a newborn. Clothes are brought into the home and put on the newborn. A baby bed or crib may be brought into the home, along with sheets and perhaps devices to keep baby's attention or make her or him sleepy, such as mobiles and soft music playing on a sound system. There may be some sort of an electronic baby monitoring system in place so parents can attend to other activities, yet stay connected to baby, in a virtual sense, from elsewhere in the house (representing a form of virtual mobility of a midlife self in order to address a need to hear the status of the baby). In large part, newborns are reliant on the motility skills and access of others to fulfill many of their needs. Self-directed, potential mobility, in the form of crawling and walking, is a little further down the road. Yet a newborn is mobile with others through mobile with arrangements. Parents may take a newborn in a car, public transit, or other means, to visit a doctor, friends, or family. Some parents speak of late night car journeys around neighborhoods as a way of helping baby fall asleep. There may be stroller or carriage rides originating from home, or in association with a journey by car, bicycle or public transit.

Brookhiser (2008, p. 62) described living in a New York City neighborhood close to a welfare office and a nursery school, and how the parked and moving strollers congest a sidewalk for those walking by. Brookhiser raised the issue of who gets to be the passenger of a stroller, taking issue not with baby passengers but with kids who could be walking. We see children who, in Brookhiser's view, could potentially be walking, but the mobile other has made choices for baby, or youngster, that keep them within a particular mobility technology, in this case in the form of a stroller, in part to address the needs and concerns of the parent or nanny. Here can be seen competing mobility action chains of different stages of the life course, attempting to utilize the same setting at the same time. Seen as mobility action chains, the connections between the newborn and needs are highly contingent on others such as family and friends (through mobile with and mobile other arrangements) who may or may not have competing needs that alter the form of particular mobility action chains. To a great extent when mobility action chains related to newborns are actualized, their precise configuration is left to others.

Youth

To date, a relatively small number of journal articles within the mobilities turn or children's geographies has focused on children's or young people's mobility (e.g., Barker et al., 2009). During youth the body changes dramatically. One learns to walk and likely ride a bicycle, for example (Jensen, 2007). Gehl (2010, p. 33) argued that one of the most memorable moments in life is the day a child is able to stand and start walking and that at this stage "life is about to start in earnest." Here Barker and colleagues (Barker et al., 2009) provided an example of this ongoing evolution, including shifting mobilities and immobilities, and the relationship between a youth and parents:

> parents' attitudes toward and practices around their offspring in cars enable youngsters to be configured as *child* car passengers—profoundly aged subjects whose experiences of car travel are de-mobilised in particular ways. On the other hand, they bespeak the multiple temporal mobilities—and *immobilities*—that undergird particular lifecourse stages. (p. 5)

Here one can see possible mobility action chains in the form of a mobile with, whereby a parent drives a car and a child is a passenger in the car (see, for example, Barker, 2009). If a car journey is required in order to connect with needs, then a youth is relying on the motility of their parent for their own motility. Based on their review of other scholars' work, Barker et al. (2009) found that children from higher social class families often have less independent mobility, with higher levels of "adult-escorted spatial mobility." The ability to drive a car rests outside the life of early youth. A legal framework, part of an operational script within the jurisdiction wherein a youth resides, will set out a threshold of characteristics, including age and skills, as to who can enter the world of car drivers and full citizenship in automobility. (For an example of an operational script involving youth and the complex relationship between youth, their parents, and the operator of a technological system in the form of a merry-go-round, see Goffman, 1975/2009.)

Thomsen (2005) noted that some of the factors contributing to children's increased automobilization include a rise in car ownership, local school closures, and a lower average age of school children. It is also suggested that parents are less positive towards children riding their bicycles to school due in part to car traffic, road safety campaigns, and media reports, raising concerns about independent mobility by children. Here we see the potential for parents to influence early independent forms of mobility action chains, which may lead to constricted motility. Barker et al. (2009) found that children are not simply passive passengers of a car, but, rather, have potential agency in dynamic and flexible ways. In this sense while a child may not be the main conductor of a particular mobility action chain configuration, he or she may influence its configuration.

Mikkelsen and Christensen (2009) found that children's mobility dependence or independence is not a fixed status as actors move in and out of dependency relationships. Once reaching the age when youth can be tested for a driver's license, some youth may hold a license while others may not, creating potential stratification within those in one stage of the life course. Dreyfus and Spinosa (2003, p. 321) provided the example of a group of teenagers looking to find their way to a music shop, leaning on one within the group who has access to a car and holds the required driving skills necessary to make their way. Without one of the group holding driving skills and having access to a car, the motility of this particular mobility action chain could collapse for the entire group. Perhaps through some reconfiguration, such as finding another friend with a driver's license and access to a car, a reconfigured chain would be established, bringing about some level of motility. Through youth one can see motility through mobile other, mobile with, and early signs of relatively independent arrangements. One can also see forms of negotiation and potentially competing needs, in how chains that bring together arrangements of a self with others, are brought about. The networked self, commencing life, naked, as an immobile, predominantly with others delivering needs, appears to expand and gain agency along available extensions through these various arrangements, in order to address needs.

Midlife

Skills and resources ramp up through the earlier stages of life. But it is in midlife that the networked self may be reaching the pinnacle of motility. The transition for some from being students in youth to becoming employed workers (possibly with families) in midlife, can bring on new mobility patterns and complex mobility action chains as they address their own needs, as well as those of their families and others. This can involve not only mobility between home and work, but multiple homes and multiple work sites, such as when the members of a couple live in different cities (Schneider & Limmer, 2008). Lassen (2006) provided the example of Hewlett-Packard employees who took on an average of 3.8 international trips per year, covering 17,000 km with 94% of international trips being by airplane. Kesselring and Vogl (2008) provided the example of distances traversed and the associated complexities as part of one individual's work and personal life mobilities. During these work trips, everyday, family-related mobile with and mobile other arrangements may have to be put off or reconfigured.

During midlife one may be a parent playing the role of mobile other and mobile with for the benefit of newborns and children, such as driving them to school (see, for example, Noy, 2009). One may also be the grown children and mobile other or mobile with of parents who are experiencing diminished mobility (Coughlin, 2001). In such a case the ample motility of the midlife self is being used to assist older parents, with fraying mobility action chains and degrading individual motility. In addition, as an employed worker, a midlife self may be offering services that are delivered to those with lesser mobility. For example, Ferguson (2009) has investigated the ways that mobility is integral to the role of social workers in addressing the needs of their clients, who may be older adults or children. As an employee of a particular organization we may be directly or indirectly involved in moving goods and information, physically and/or virtually, from one place to another to address the needs of clients or others. A midlife self may have the motility, and what Urry (2007) has described as *obligation*, to visit with those who do not have the degree of motility available earlier in life. From McEwan (2005) we draw the following example:

> driving west in light traffic, Perowne finds he's feeling better about seeing his mother. He knows the routine well enough.... Being with her isn't so difficult. The hard part is when he comes away...when the woman she once was haunts him as he stands by the front door and leans down to kiss her goodbye. That's when he feels he's betraying her, leaving her behind in her shrunken life, sneaking away to the riches, the secret hoard of his own existence. Despite the guilt, he can't deny the little lift he feels, the lightness in his step when he turns his back and walks away from the old people's place and takes his car keys from his pocket and embraces the freedoms that can't be hers. (p. 153)

Here one can see the striking contrast between the motility available to two different generations at one precise location and moment in time. Whereas in the past there may have been available mobility action chains that could be actualized by either the grown child or the elderly parent in order to spend time together, now, based on changing motility of the elderly parent, the grown child finds that he or she is obliged to actualize their motility and undertake a particular mobility action chain in order to create a connection. What is unmentioned here, yet also likely in place, is the staff of the home who address the needs of the late-life self, and who represent a form of mobile other.

Phillips and Bernard (2008) described an example of the mobility of a daughter in midlife, who juggles the demands of part-time work with the responsibilities for caring for her elderly mother, and the resentment she feels towards her sisters who, while willing to help, live too far away to deliver effective, day-to-day care. The mobility action chains, including their mobile with and mobile other arrangements, related to visiting and assisting her mother, are likely different from her mobility action chains from a time where her mother did not require assistance. Through midlife we see independent motility soar. With new life obligations such as partners, children, work, and parents, the midlife self is faced with potentially competing elements demanding their time, and seeking to borrow or utilize their motility via mobile with and mobile other arrangements to address competing needs.

Late-life

In late-life motility may be diminishing, in a sense being stripped away, taking one back towards the naked self, as skills and access are potentially in jeopardy. For example, we can consider those experiencing late-life in lower density, built environments where there is a strong reliance on automobility. In this regard Rosenbloom (2003) raised alarm bells at the large number of seniors living and aging in lower density, suburban settings in the United States, and questioned how they will connect with needs in the future. Similarly, from a Canadian perspective, Hodge (2008) noted that the spatial arrangements of dwellings in a community can exert environmental pressure on seniors by impacting the extent of possible activity patterns (i.e., mobility action chains) in environments that may be highly dependent on cars.

In a qualitative research study on seniors living in suburban Boston one senior stated "Everything out there is moving. If you're not a part of it... it's tough" (Coughlin, 2001, p. 5). In the same study a suburban non-driving senior stated "not driving...you become a prisoner. I have to depend on other people" (Coughlin, 2001, p. 6). Coughlin stated: "Driving is so much a part of the American identity.... Not being able to drive is not just giving up a certain mobility mode, but also changing how we define independence and freedom" (cited in Stromberg, 2007, p. 9). For some, it may be that walking degrades earlier than the ability to drive a car, and driving becomes more important as a means of maintaining connections (Siren & Hakamies-Blomqvist, 2009). American architects and town planners, Duany, Plater-Zyberk, and Speck (2000) aptly summarized the sense of living in a built environment given a design script centered around young family mobility, which potentially no longer meets one's late-life needs and the increased mobility challenges:

> as soon as they lose their driver's licenses, the location of that house puts them out of reach of their physical and social needs. They become in effect nonviable members of society. Unless they are wealthy enough to have a chauffeur, or are willing to burden a relative, they have no choice but to re-retire into a specialized home for the elderly...they spend the rest of their days quarantined with their fellow nonviable members of society. (pp. 122-123)

Duany et al. (2000) described a scenario where the skills required to hold motility in the form of a particular mobility action chain, within a particular extension, is greatly diminished. Some will be able to undertake coping strategies that avoid driving in certain circumstances where skills may now be degraded, in an attempt to hold onto established mobility action chains, now perhaps through arrangements with others, or configure new mobility action chains that do not involve a car (Donorfio, D'Ambrosio, Coughlin, & Mohyde, 2009; Kostyniuk & Shope, 1998; Rudman, Friedland, Chipman, & Sciortino, 2006).

One can see needs and their fulfillment through connections potentially at risk as the ability to drive a car diminishes, or is lost completely, in a built environment that requires utilizing a car as a driver or passenger. Re-retiring to a "specialized home" would greatly reconfigure traditional mobility action chains towards a higher proportion of mobile others assisting with connecting the late-life self with needs. From a life course perspective, and reflecting on Hall's thoughts on considering extensions that are suited to some while being ill suited to others, one can see those in late-life, newborns, and youth as holding similar challenges within a built environment designed predominantly for automobility. Considered in isolation of their possible mobile with or mobile other arrangements, they may be relatively immobile, or hold low motility, and thus lack the capacity to address needs that have engendered mobility action chain options requiring the use of a car. From his research on Boston seniors, Coughlin (2001) noted that seniors may ask friends and grown children for a ride, but there is often a degree of uneasiness in making the request.

Pooley et al. (2005) found that in some instances seniors will restrict or alter their mobility due to safety concerns, even though skills are in place. This in turn creates reconfigured mobility action chains. This may be a situation where a need for safety is given priority over other needs, but additional research is required to clarify this. Pooley et al. (2005, p. 14) also found that even for those who are housebound, mobility is a crucial part of everyday life. Martin-Matthews (2007) has explored homecare (what can be seen as a form of mobile other) from the perspective of the older person receiving care from their family and formal care providers. There may be discomfort with "strangers" in a home, yet the trade-off is the possibility of not having to reconfigure where "home" is. Reconfiguring home creates the necessity of establishing new mobility action chains in order to connect the self with needs. As one 79-year-old, female care receiver described it:

> Having a homemaker helps keep people in their own homes. It's worth it—the training new people, the constant revolving door with new people coming into my home and my having to explain everything to each one all over again. I don't like that, but I know what the alternative is. (Martin-Matthews, 2007, p. 236)

End of Life

Lastly is end of life. Each unique, networked self has taken a unique life journey, many with individual motility rising through early life, peaking in midlife, and diminishing slowly or abruptly in late-life. Similar to the beginning of life there is often a degree of unpredictability to the exact timing of the end of life. For some, the end comes quite suddenly with a rush of mobile others and mobile withs attempting to offer life-saving emergency response services. For some, there may be an extended period where one is living housebound (see, for example, Schwartz, 1996) in their own home, or in an institutional setting (see Goffman, 1961 regarding "total institutions" and how they have to address the needs of those who live in them), with staff and family coming to support end-of-life needs through mobile other arrangements.

Conclusion

This chapter began by presenting a theoretical frame that can be used as a lens through which one can consider motility of the networked self across the life course. Glimpses of motility at various stages of the life course were presented. By considering mobility technologies and mobility infrastructures utilized by the self, and related skills and access, one can see how available and utilized extensions continue to change across the life course, at some stages growing while at other stages degrading. Through early life the self is reliant on mobile with and mobile other arrangements for its motility. In midlife the self has ample motility that can at times be offered to newborn, youth, and late-life selves as a means of enhancing their motility. Somewhat like a reverse mirror image of early life, the late-life self may build up mobile with and mobile other arrangements, in an attempt to sustain motility, as individual motility degrades. By considering mobility action chains, one can see how unique and particular connections are configured and reconfigured across the life course, including independent, mobile with, and mobile other arrangements that attempt to keep the self connected to needs. When all brought together, one can see the unique, evolving configurations, plasticity, and malleability of the networked self across the life course.

Utilizing the lens presented here, and the glimpses of motility considered, one could consider this a preliminary step towards a deeper understanding of motility, where one could go further still in distilling the complexities of motility of the networked self across the life course. One possible course forward is to undertake detailed consideration of the multiple generations of a particular family, to see how motility of the self grows and degrades and how configured and reconfigured mobility action chains attempt to build or retain motility. One could also examine how motility can rise and fall through mobile with and mobile other arrangements with other family members. Such an approach could include GPS data where one could see mobile with arrangements exist, where two or more selves bring their mobility action chains together for some moment in time and space and then part ways again. At the same time there could be an attempt to grasp a deeper understanding of the complementary and competing needs of the self and those with whom the self attempts to create arrangements, to see how this relates to the formation of successful or unsuccessful mobility action chains. Yet another possible course is to consider a specific mobility setting and investigate how this setting is utilized, or not utilized, tracing out unique mobility action chains, including independent, mobile with, and mobile other arrangements, by those at different stages of the life course. With such an approach one could consider in some detail the design and operational scripts which may encourage or discourage particular mobility action chain configurations of the self at various stages of the life course.

References

Adey, P. (2010). *Mobility*. Abingdon, UK: Routledge.

Akrich, M. (1992). The de-scription of technical objects. In W. E. Bijker & J. Law (Eds.), *Shaping technology / building society: Studies in sociotechnical change* (pp. 205-224). Cambridge, MA: MIT Press.

Barker, J. (2009). 'Driven to distraction?': Children's experiences of car travel. *Mobilities*, *4*(1), 59-76.

Barker, J., Kraftl, P., Horton, J., & Tucker, F. (2009). The road less travelled—New directions in children's and young people's mobility. *Mobilities, 4*(1), 1-10.

Bijker, W. E., & Law, J. (1992). What next? Technology, theory, and method. In W. E. Bijker & J. Law (Eds.), *Shaping technology / building society: Studies in sociotechnical change* (pp. 201-203). Cambridge, MA: MIT Press.

Brookhiser, R. (2008, July 14). City desk: Props to the pram—Richard Brookhiser praises the stroller, *National Review, 60*(13), 62.

Burgess, E. W. (1967). The growth of the city: An introduction to a research project. In R. E. Park, E. W. Burgess, & R. D. McKenzie (Eds.), *The city* (pp. 47-62). Chicago, IL: The University of Chicago Press.

Canzler, W., Kaufmann, V., & Kesselring, S. (2008). Tracing mobilities: An introduction. In W. Canzler, V. Kaufmann, & S. Kesselring. (Eds.), *Tracing mobilities: Towards a cosmopolitan perspective* (pp. 1-10). Aldershot, UK: Ashgate.

Cavell, R. (2002). *McLuhan in space: A cultural geography.* Toronto, Canada: University of Toronto Press.

Coughlin, J. (2001). *Transportation and older persons: Perceptions and preferences: A report on focus groups.* Washington, DC: American Association of Retired Persons (AARP).

Donorfio, L. K. M., D'Ambrosio, L. A., Coughlin, J. F., & Mohyde, M. (2009). To drive or not to drive, that isn't the question—The meaning of self-regulation among older drivers. *Journal of Safety Research, 40*(3), 221-226.

Dreyfus, H. L., & Spinosa, C. (2003). Heidegger and Borgmann on how to affirm technology. In R. C. Scharff & V. Dusek (Eds.), *Philosophy of technology: The technological condition—An anthology* (pp. 315-326). Malden, MA: Blackwell Publishing.

Duany, A., Plater-Zyberk, E., & Speck, J. (2000). *Suburban nation: The rise of sprawl and the decline of the American dream.* New York, NY: North Point Press.

Ferguson, H. (2009). Driven to care: The car, automobility and social work. *Mobilities, 4*(2), 275-293.

Gehl, J. (2010). *Cities for people.* Washington, DC: Island Press.

Goffman, E. (1959). *The presentation of self in everyday life.* New York, NY: Anchor Books.

Goffman, E. (1961). *Asylums: Essays on the social situation of mental patients and other inmates.* New York, NY: Anchor Books.

Goffman, E. (2009). Role distance. In D. Brissett & C. Edgley (Eds.), *Life as theater: A dramaturgical sourcebook* (pp. 101-111). New Brunswick, NJ: Transaction Publishers.

Hall, E. T. (1983). *The dance of life: The other dimension of time.* New York, NY: Anchor Books.

Hall, E. T. (1989). *Beyond culture.* New York, NY: Anchor Books.

Hall, E. T. (1990a). *The hidden dimension.* New York, NY: Anchor Books.

Hall, E. T. (1990b). *The silent language.* New York, NY: Anchor Books.

Hodge, G. (2008). *The geography of aging: Preparing communities for the surge in seniors.* Montreal, Canada: McGill-Queen's University Press.

Ihde, D. (1990). *Technology and the lifeworld: From garden to earth.* Bloomington, IN: Indiana University Press.

Ihde, D. (2002). *Bodies in technology.* Minneapolis, MN: University of Minnesota Press.

Ihde, D. (2010). *Heidegger's technologies: Postphenomenological perspectives.* New York, NY: Fordham University Press.

Jacobs, J. (1992). *The death and life of great American cities.* New York, NY: Vintage Books.

Jensen, O. B. (2007, August). *Biking in the land of the car—Clashes of mobility cultures in the USA*. Unpublished paper presented at Trafikdage på Aalborg Universitet, Aalborg, Denmark.

Jensen, O. B. (2009a). Flows of meaning, cultures of movement—Urban mobility as meaningful everyday life practice. *Mobilities, 4*(1), 139–158.

Jensen, O. B. (2009b). Foreword: Mobilities as culture. In P. Vannini (Ed.), 2009 *The cultures of alternative mobilities: Routes less travelled* (pp. xv–xix). Farnham, UK: Ashgate.

Jensen, O. B. (2010). Erving Goffman and everyday life mobility. In M. H. Jacobsen (Ed.), *The contemporary Goffman* (pp. 333–351). New York, NY: Routledge.

Kaufmann, V. (2002). *Re-thinking mobility: Contemporary sociology*. Aldershot, UK: Ashgate.

Kellerman, A. (2006). *Personal mobilities*. London, UK: Routledge.

Kesselring, S., & Vogl, G. (2008). Networks, scapes and flows—Mobility pioneers between first and second modernity. In W. Canzler, V. Kaufmann, & S. Kesselring (Eds.), *Tracing mobilities: Towards a cosmopolitan perspective* (pp. 163–179). Aldershot, UK: Ashgate.

Kostyniuk, L. P., & Shope, J. T. (1998). *Reduction and cessation of driving among older drivers: Focus groups*. Ann Arbor, MI: University of Michigan, Transportation Research Institute.

Lassen, C. (2006). Aeromobility and work, *Environment and Planning A, 38*(2), 301–312.

Lassen, C., & Jensen, O. B. (2004). Den globale bus—Om arbejdsrejsers betrydning i hverdagslivet. In M. H. Jacobsen & J. Tonboe (Eds.), *Arbejdssamfundet* (pp. 241–280). København (Copenhagen), Denmark: Hands Reitzels Furlag.

Latour, B. (1992). Where are the missing masses?: The sociology of a few mundane artifacts. In W. E. Bijker & J. Law (Eds.), *Shaping technology / building society: Studies in sociotechnical change* (pp. 225–258). Cambridge, MA: MIT Press.

Malabou, C. (2008). *What should we do with our brain?* New York, NY: Fordham University Press.

Marshall, V. W., & Mueller, M. M. (2003). Theoretical roots of the life-course perspective. In W. R. Heinz & V. W. Marshall (Eds.), *Social dynamics of the life course: Transitions, institutions, and interrelations* (pp. 3–32). New York, NY: Aldine de Gruyter.

Martin-Matthews, A. (2007). Situating 'home' at the nexus of the public and private spheres: Ageing, gender and home support work in Canada, *Current Sociology, 55*(2), 229–249.

McEwan, I. (2005). *Saturday*. Toronto, Canada: Alfred A. Knopf.

McLuhan, M. (1994). *Understanding media: The extensions of man*. Cambridge, MA: MIT Press.

Mikkelsen, M. R., & Christensen, P. (2009). Is children's independent mobility really independent? A study of children's mobility combining ethnography and GPS/mobile phone technologies. *Mobilities, 4*(1), 37–58.

Mitchell, W. J. (2004). *Me++: The cyborg self and the networked city*. Cambridge, MA: MIT Press.

Mumford, L. (1963). *Technics and civilization*. San Diego, CA: Harvest Books.

Noy, C. (2009). On driving a car and being a family: An autoethnography. In P. Vannini (Ed.), *Material culture and technology in everyday life* (pp. 101–113). New York, NY: Peter Lang.

Phillips, J. E., & Bernard, M. (2008). Work and care: Blurring the boundaries of space, place, time, and distance. In A. Martin-Matthews & J. E. Phillips (Eds.), *Aging and caring at the intersection of work and home life: Blurring the boundaries* (pp. 85–105). New York, NY: Lawrence Erlbaum.

Pooley, C. G., Turnball, J., & Adams, M. (2005). *A mobile century?: Changes in everyday mobility in Britain in the twentieth century*. Aldershot, UK: Ashgate.

Richardson, T., & Jensen, O. B. (2008). How mobility systems produce inequality: Making mobile subject types on the Bangkok Sky Train. *Built Environment, 34*(2), 218–231.

Rosenbloom, S. (2003). *The mobility needs of older Americans: Implications for transportation reauthorization.* The Brookings Institution Series on Transportation Reform, Center on Urban and Metropolitan Policy. Washington, DC: Brookings Institution.

Rudman, D. L., Friedland, J., Chipman, M., & Sciortino, P. (2006). Holding on and letting go: The perspectives of pre-seniors and seniors on driving self-regulation in later life. *Canadian Journal on Aging, 25*(1), 65–76.

Schneider, N. F., & Limmer, R. (2008). Job mobility and living arrangements. In W. Canzler, V. Kaufmann, & S. Kesselring (Eds.), *Tracing mobilities: Towards a cosmopolitan perspective* (pp. 119–139). Aldershot, UK: Ashgate.

Schwartz, M. (1996). *Morrie: In his own words.* New York, NY: Walker and Company.

Shane, D. G. (2005). *Recombinant urbanism: Conceptual modelling in architecture, urban design and city theory.* Chichester, UK: Wiley.

Simmel, G. (1971). *Georg Simmel: On individuality and social forms: Selected writings.* Chicago, IL: University of Chicago Press.

Siren, A., & Hakamies-Blomqvist, L. (2009). Mobility and well-being in old age. *Topics in Geriatric Rehabilitation, 25*(1), 3–11.

Stromberg, M. (2007). Growing old in a car-centric world. *Planning, 73*(10), 6–11.

Thomsen, T. U. (2005). Parents' construction of traffic safety: Children's independent mobility at risk? In T. U. Thomsen, L. D. Nielsen, & H. Gudmundsson (Eds.), *Social perspectives on mobility* (pp. 11–28). Aldershot, UK: Ashgate.

Turner, J. S. (2000). *The extended organism: The physiology of animal-built structures.* Cambridge, MA: Harvard University Press.

Urry, J. (2007). *Mobilities.* Cambridge, UK: Polity Press.

Urry, J. (2008). Moving on the mobility turn. In W. Canzler, V. Kaufmann, & S. Kesselring (Eds.), *Tracing mobilities: Towards a cosmopolitan perspective* (pp. 13–23). Aldershot, UK: Ashgate.

9

Seniors, Cell Phones, and Tactical Restriction

Kim Sawchuk and Barbara Crow

Over the past decade there has been a growing interest in the concept of mobility concomitant with the proliferation of mobile media devices for communicating. Much of this work has centered on the cellular telephone (see Caron & Caronia, 2007; Castells, Fernández-Ardèvol, Linchuan Qiu, & Sey, 2006; Cresswell, 2006; Goggin, 2006; Gow & Smith, 2006; Ito, Okabe, & Matsuda, 2005; Middleton & Cukier, 2006; Urry, 2007), with a particular interest in how young and middle-aged users engage with these technologies (see Brants & Frissen, 2005; Caron & Caronia, 2007; Haddon, 2005; Horst & Miller, 2006; Katz & Aakhus, 2002; Wajcman, Bittman, & Brown, 2008). Except for a handful of articles, largely absent from the communication and cultural studies literature on mobile media and cell phones are the practices and perceptions of anyone over sixty years of age.[1]

Two statistical trends are noteworthy in relation to this relative absence. First, the number of people in Canada who will be sixty-five and over is predicted to double from 4.2 million at present to 9.8 million by 2038.[2] Second, the mobile phone, or cell phone, is rapidly displacing the landline telephone for person-to-person communications and increasing in popularity worldwide. Canada has had lower and slower rates of mobile telephone adoption than elsewhere in the world, but this is shifting.[3] On average, 75% of Canadians now own a cell phone, a steady increase since 1997. The lowest rates of ownership are in Quebec, and within this province the lowest rate of diffusion is amongst those who are aged fifty-five and over (Canadian Wireless Telecommunications Association (CWTA), 2008).

As researchers on the use of mobile phones by this population note, older cell phone owners tend to use it "instrumentally" and judiciously, for reasons of personal safety, rather than as a means to engage in perpetual contact with friends or even with relatives (Ling, 2004). Data collected by the wireless telecommunications industry as well as research in the field of human-computer-interaction indicates that when age cohorts use their phones there are very specific generational patterns. Those who are sixty and over are more likely to restrict their use of the cell phone for emergency calls only, are more likely to call family than friends (Conci, Pianese, & Zancarano, 2009; Wong, Thwaites, & Khong, 2008), and are less likely to use any of the multifunctional capabilities of the phone (Kurniawan, 2008; Mohd, Hazrina, & Nazean, 2008). The younger the user, the more likely it is that they will send a text message, take a picture, download content, and send or receive multimedia messages (Canadian Wireless Telecommunications Association [CWTA], 2008; Lenhart, 2010). The reasons given for these differences tend to focus on the design features in relationship to the cognitive abilities of individual senior users.

In summarizing the current state of the debate on the mobile phone use amongst the elderly, Rich Ling (2008) asks the following set of questions: "Are the elderly victims of poor design and exclusion or are they agents in their own isolation? Should the policy be to include the elderly in every possible digital development? Is their quasi self-chosen aloofness a result of poor design? Will the gap eventually disappear?" (p. 338). Ling's questions are pertinent, however, his concern *for* the elderly *from* the perspective of the "young" user is indicative of a larger set of methodological problems in the literature on mobile telephony. What is absent in most of these articles are the voices of seniors.

In this chapter, we argue against the typical, oft-times pejorative framing of seniors in the literature as "hesitant" (Charness & Boot, 2009; Ling, 2008) or "fearful" users of technology (Kurniawan, 2008) who are either incapable or unable to access the multifunctionality of the mobile phone. We do not wish to underplay the very real problems that may be faced by this group because of design issues, including screens that are too small, buttons that are awkward, or menu systems that are incomprehensibly complex, all valid concerns to address with respect to access. Instead, our research begins with what seniors who use the cell phone do with their cell phones.

Our interest in this approach is based on over 40 group interviews with over 200 participants, and two years of observational experience with those who are sixty-five and over across Canada. The average age of our participants, to date, is seventy-one. In our discussions we have asked these seniors how long they have had a phone, how they acquired it, what they use it for, when they use it, and how the cell phone fits within a broader pattern of communications. We are still analyzing these data, but our thinking is inspired by cultural approaches to cell phone studies, including the writings of Virpi Oksman (2006), Gerard Goggin (2006), Heather Horst and Daniel Miller (2006), and Sally Wyatt (2005). In particular, we are influenced by Oksman and Wyatt, who reconsider the perspective of the user, advocating a continuum of use and an analysis of the conditions of usage within particular, local contexts.

While we borrow from recent ethnographic approaches to understanding the world view of our constituency (Counts & Counts, 2001), as we further contend, the omission of their mobile practices reveals a bias in how researchers approach the study of "users" and "non-users" (Wyatt, 2005), where there is a presupposition that frequency is a sign of engagement. What if their restrictive use-practices are understood as an informed decision or a valid, tactical response to broader systemic issues? What if their limited use of the cell phone is situated within an entire media ecology of telecommunications? Studies of usage need to understand the restriction of functions and services by users not merely as aloofness or inability, but as a deliberate set of interrelated actions, including tactical restrictions.

Here we focus on a specific group: ten female quilters, between the ages of 65 and 82, living in a rural area near Toronto, whom we refer to as the Lakeside Quilters. What is unique about the quilters is that they are a community of female friends whose lives are involved in a shared passion-quilting. Their commitment to their craft demands a high level of manual dexterity and puts them into contact with other digital technologies, including complex sewing machines that require a degree of programming. Our attention to the experiences of senior women living in a rural region is deliberate in the context of a culture that values youth and beauty, and where the implications of ageing, economically and socially, are different for men and for women (Riggs, 1998, 2002, 2004). Women generally live longer, and the price of this longevity is also a significant statistical gap in income between men and women (Canadian Association of Social Workers, 2004; Conference Board of Canada, 2009).

We poach the term *tactics* from Michel de Certeau's (1984) *The Practice of Everyday Life*, explicating the myriad ways that those in positions of subordination negotiate disciplinary systems of exclusion or indifference. De Certeau's analysis foregrounds everyday practices such as walking and cooking and furnishes a cogent, yet wonderfully mundane set of examples of why these everyday engagements matter. As Brian Morris (2004) has written, de Certeau's text, revisited from the perspective of urban studies, "might provoke new kinds of engagements for scholars—engagements oriented more firmly to the ways in which bodies, subjects and built environments are interlinked and enmeshed" (p. 676). As a contribution to the field of cultural and media studies that focuses on the practices of everyday life as a "form of enunciation" (Morris, 2004, p. 677), de Certeau asks us to consider how a variety of enunciative actions, or practices, actualize possibilities within material conditions of constraint and official orders that contain a variety of normative assumptions.

De Certeau's discussion turns on a distinction between strategies and tactics reminding us that political, economic, and scientific rationalities are constructed on a "strategic" model that is a "calculus of force-relationships which becomes possible when a subject of will and power (a proprietor, an enterprise, a city, a scientific institution) can be isolated from an 'environment.' A strategy assumes a place that can be circumscribed as proper and thus serve as the basis for generating relations with an exterior distinct from it (competitors, adversaries, clientele, targets, or objects of research" (de Certeau, 1984, p. xix). In contrast, a tactic is a practice that inserts itself "into the other's place, fragmentarily, without taking it over in its entirety, without being able to keep it at a distance" (de Certeau, 1984, p. xix). Tactics depend on the "art of timing" and the ability to watch for opportunities that are seized on the fly. Tactical usage involves an appropriation of what is not rightfully intended, or inscribed, as yours. But tactics are never simplistically opposed to strategies. For de Certeau, these tactical manoeuvres and collective narrative constructions between people in the context of everyday life underscore the moral choices, the aesthetic performances, and the micropolitical negotiations with strategies of disciplinary exclusion.

Within the field of mobility studies, the distinction between strategies and tactics is a recurrent one, particularly for those studying mobile media practice (O'Brien, 2009; Pierson, Mante-Meijer, Loos, & Sapio, 2008; Wilken, 2005). Yet while the distinction between strategies of power and tactics of resistance to both technology and urban space is a *leitmotif* in the mobilities literature, the studies on tactical appropriations of mobile media technologies focus on gender (Gregg, 2004), youth (O' Brien, 2009; Wilken, 2005), or simply generalize and speculate on the activities of unspecified users (Pierson et al., 2008). "Tactics" are ways to resist or negotiate the prescriptive norms that govern systems and structures that devalue the aged and the processes of ageing.

The concept of "tactical restriction" is a way to think about the practices of our senior users to do justice to their attempts to work through and around conditions of constraint and unequal relations of power. Framing their practices as *tactical* valorizes their experience with mobile media culture in ways that differ from how they are typically depicted in the literature: in unintentionally patronizing terms that oft-times compares them to a more youthful, exuberant, ideal counterpart.

Engaging in a discursive analysis of what Potter and Wetherell (1987) have called "interpretive repertoires" we explore four ways that the Lakeside Quilters engage in tactical media usage. First, we look at the role that the cell phone plays in their preparations for emergency situations. We examine the economic and cultural conditions that lead this group to limit their cell phone consumption. We then consider, briefly, their understanding of the need to manage time, but also the options exercised in terms of other media, placing their use of the cell phone in the context of a range of media options. Finally, we address a key issue: some of the tactics, such as performance and disguise, used to negotiate ageism in the retail setting.

Tactical Restrictions: Emergency Uses

Ask a senior why they have a cell phone and the most frequent response is "for emergency purposes." This motivation to buy a cell phone is one that recurs with astonishing frequency across much of the existing literature on seniors and cells, and is echoed over and over again in our interviews. The cell phone is "handy" for use in *exceptional* circumstances. However, while the explanation of "the emergency" is often cited as the number one motivating factor for seniors to purchase a cell phone, how an emergency is conceived is rarely articulated. As Oksman (2006) suggests in her study of Finnish seniors and their cell phone practices, while the rationale and word choice may be the same, the idea of what constitutes an actual emergency situation is highly contextual. As she asserts, "knowing the actual use contexts and user experiences of technology among different generations in their daily life can provide important insights on how to improve design and services associated with technologies" (p. 4). And so we turn to our interview material, and the ways that the Lakeside Quilters differentiate categories of "emergency" within their local context.

The Lakeside Quilters distinguished between major emergencies such as health-related matters, car trouble, or an accident, and minor emergencies such as being late, needing to pick up something from the store, or using the cell, as another senior commented, "as a spousal locator" in an overly large store. What is important to understand is the specificity of the term *emergency* in this rural context. Unlike our urban users, such as those living in Toronto, discussions of emergency were often entangled with the term, "personal safety," and vulnerability on the street. In the case of the Lakeside Quilters, emergency was connected either to driving or to health-related matters.

For this group of rural women mobility primarily meant having access to a car. When questioned about the term, *mobility*, the car was mentioned more frequently than the cell phone, and the loss of a driving license was seen as one of the greatest threats to their sense of mobility: "I think the hardest thing is the idea that I may have to give up the car" (Judith, 77). These seniors depend on automobile travel in all four seasons, including Canadian winters, when highway driving can be hazardous. Because of the possibility of inclement weather a phone in the car was seen as critical. "You know, to be able to go out, and get in the car and go.... I dread snowstorms, because I promised the kids I wouldn't drive in the snowstorms or icy roads. And then...your mobility is gone" (Doris, 69). In these instances, it was also noted that phone booths are, in rural regions, a rarity. "A phone booth when you really need to get in touch? That's why I have a cell phone in the car, and it's there for me when I'm in trouble" (Jean, 74). The cell phone was imagined as potentially useful in an emergency situation, such as a road accident. While the phone was connected to their ability to be mobile and stay mobile in general, there was solid agreement that speaking on a cell phone while driving was a no-no.

Given the age of our population, the phone was also seen as a useful device for maintaining ongoing contact with family members in case of a health-related emergency. One participant explained:

> I had a stroke nine years ago, and when I was doing some of the testing, I was all wired up with a whole bunch of wires and my car stopped. And I sat on the 115 for ages, and nobody came. Anyway, someone did finally pick me up and take me to a telephone. So my husband insisted I have a phone there for emergencies. (Ruth, 70)

In other instances, the phone was purchased to keep in contact with family and friends during an emergency: "We got ours for emergencies. My husband was going in for bypass surgery, and I just knew that I had to keep in contact with children and family and all, and that's the route we went" (Evelyn, 61). The cell phone is a tool for "emergency" assistance, a kind of "insurance" if something goes wrong, rather than a tool used for habitual contact with family and friends. Even if used on a daily basis, the contact would be infrequent, and long calls would be made from the home.

Economic Tactics to Maximize Use

This primary restriction of the mobile phone to emergency usage is related to the economy and care in the management of one's goods and finances. As we know from statistical data on seniors and poverty, one's income drops significantly after retirement. In 2008, married couples earned a combined income of $80,500 (on average, after taxes), while elderly married couples earned a combined income of $53,500 (Statistics Canada, 2010). In Canada, cell phone service contracts are onerous and costs tend to be high relative to the services provided (Wireless North, 2010). As report after report notes, Canada is the only country where users are charged for long distance calls in another area. Thus, seniors have to take care of not only how they use cell phone service, but where they use it and how much they use it.

These costs did not go unnoticed by the quilters or by other seniors who have begun to contact us to participate in the project to, in the words of one email writer, "get the message out to the companies." In the context of sharing a car between family members, one notable feature of the quilter's money-saving tactics in cell phone use was the sharing of a single device between two people: "We have one my husband and I share, and it's basically probably used for a little bit more than emergency, but we only pay $10 a month and it's a good insurance policy for when you're travelling..." (Carol, 75).

This discursive fragment also addresses an important presupposition embedded in the collection of statistical data on cell phones: the idea that *individual* ownership of a phone is seen as equivalent to access. Except for one person, the seniors in both our individual interviews and discussion groups all had *access* to a cell phone, but they did not necessarily have individual ownership of the device. Many seniors *share* one phone within their household, which in this case meant between husband and wife. In other words, the pattern of use follows a household model of understanding the ownership of technology, rather than a more individualistic model that tethers the cell phone, and individual number, to a single person.

In sharing the phone between two people, one sees the tactical practices of this group in action operating within a structural set of economic conditions imposed by Canadian telecommunications companies on consumers. If our cell phone sharing couples are out together, the phone is theirs. If one is away, they take the phone (and most likely the car) with them. In this way, the phone in the hands of our thrifty seniors is not a personalized, customizable, intimate object or fashion statement, tailored to their taste. It is a "utilitarian" device. Many also commented that they did not change their phones frequently, a finding commensurate with other research.

The Lakeside Quilters were especially wary of the bundling of digital communications: cable, Internet, landline, and cell phones all together. Most had pay-as-you-go plans and kept very careful track of their minutes. Rather than seeing bundles as a cost saving, there was a concern that these bundles would make them dependent on one company for all of their telecommunication services. Many recounted struggles with companies over mistakes and unfair charges:

> I have friends for whatever reason who have kids they keep in touch with and everything and they have had huge rows with these companies. And I just switched from Bell to Cogeco and I had a big fight with them, and I don't need another fight on my hands. (Elizabeth, 64)

Stories of struggle with companies out to take their hard-earned money frequently became a part of the discussion with almost every group of the seniors we have interviewed these past two years.

For the Lakeside Quilters, the question of cost is not just about being able to afford a device or service. This group expressed a set of values towards money that is part of what they understand as a part of their generational identity. As several quilters mentioned (at least three times) "we are a generation of 'savers'" (Evelyn, 61). This sense of fiscal responsibility, care with budgeting, and economizing, is inflected in their cell phone practices, including keeping close track of the minutes available on a prepaid plan, and then using up any spare minutes on long distance, even if it was more costly than a regular landline, so that the unused minutes would not go back into the coffers of the telecommunications companies.

Indeed, this attitude towards "waste," their attention to counting pennies, checking bills, and attention to financial details could be seen as an integral part of the quilters' ethos of reusing and recycling fabric. But it is also an important part of negotiating a system of financing and payment so that it works to their advantage. And it is also a reflection of their consciousness of managing their pensions. For example, this group of users reflected on a day when there might not be a need for landlines: "But I would think that in a number of countries, people don't have landlines. And I think that probably in reality, that will eventually come, that we will all have cell phones and stuff." However, she was quick to add that in the case of living in rural Canada, "until the rates become affordable, it's prohibitive for some of us who are on, basically, fixed incomes" (Evelyn, 61).

Like many of their counterparts elsewhere in the country, this cohort preferred "pay as you go" plans because they allow for more careful monthly budgeting:

> I'm with Virgin Mobile, don't ask me how I got there, but I've always been pay-as-you-go. And then I found out, what I think is the cheapest way, that I can buy $100 and it's good for a year, and I don't know what the limit is, but I'll never get there. But I figured it was cheaper than $10 a month. (Judith, 77)

On a practical and technical side, these phones also have fewer features and functions.

The limitations of the pay-as-you-go phone provides an alternative way to understand why seniors do not engage in a wider range of mobile practices: their bargain budget phones and pay-as-you-go plans simply do not allow for it. In an extensive study of seniors living in England, for example, Sri Kurniawan (2008) noted that older people are passive users of mobile phones, experience fear of consequences of using unfamiliar technology, and most prefer design features as aids for declining functional abilities. While Kurniawan provides insights into older users' perception and use, and cost is mentioned as the most important "social factor" in this analysis, this important insight is not factored into her conclusions.

Time Management

As de Certeau writes, tactics are not only about the negotiation of space and official orderings. Tactics require users to be "on the watch for opportunities that must be seized 'on the wing'" (1984, p. xix). Time management was a recurrent theme in our discussions with the quilters. The conversation about managing costs was inflected with discussions about how to manage their time and social relations. Many commented on not feeling the need to be "on" all of the time. Several times, the quilters admitted that they preferred to call out, rather than have the others call them: "I never turn mine on, unless I'm making a call" (Ruth, 70), and "Mine's in my purse, not turned on" (Francis, 71). Others had developed strategies to work within their budgetary limits, by letting it ring, seeing who was calling before answering, and calling from a landline upon reaching a destination. Using the feature of a missed call is a strategy shared by both teenagers and seniors, who use the phone almost as a pager or adopt ways of sending messages to inform others where they will meet them later.

This was not only a cost-saving measure. Calling out, rather than calling in, allowed these women to safeguard their privacy. The concentration demanded by their quilting meant that this cohort valued selective isolation as much as connection. As one quilter stated bluntly, one of the negative sides of so much mobility and flexibility in communications was "that you can never just be by yourself" (Jean, 74). Being interrupted with constant communications was seen as a potential nuisance for both individual and social reasons.

Restricting use was related to a question they returned to repeatedly: how are cell phones changing human interaction? How is their constant presence transforming private and public space? Restaurants were favorite examples of annoying, disrespectful uses of the cell phone: "When people are in a restaurant and they're talking to somebody, for some reason, they feel they have to talk *loud* [everyone chimes in, laughter]. They are infringing upon my space" (Evelyn, 61). Another key question was whether the cell phone was a tool for enhancing social and familial communication or for impeding it.

Our quilters noted that rather than talking to each other, it was common to be interrupted by a cell phone, or texting to people who were not at the table. Many of their observations were about changes they noted in the behaviors between now and "back then," particularly with regards to children and digital technologies. Many of their questions concerned play: "Kids don't play outside" (Ruth, 70) like they used to. But while there were clear differentiations made between now and then, the comments were not only negative. There was a great deal of envy at how adept young children were at learning technology, using these devices for creative purposes, such as "to get pictures to color."

Also, many clearly relied on family, including grandchildren, for help with troubleshooting problems with both their cell phones and their computers. Yet, at the same time, one of the concerns of these quilters was that in the celebration of the new, this generation would not learn to "appreciate old things." At these moments in the conversation, the discussion of new technologies often slid into a discussion of age and ageing, pointing to an intuitive awareness of the propensity to value what is new over what is considered "old," passé, or out-of-date.

In reviewing the discussion transcripts, the intertwining of ageing with the capacity for learning new skills, and a concern with keeping up with the developments in a changing digital world, became evident. Here it is important to know that this is not because this is a group afraid of engaging with digital technologies or new learning. "I do like technology, and I just want to know how to use it with less frustration" (Jean, 74). Many of these women had been long-term members of the Lakeside Quilters. Their sewing machines used computer software that needed to be updated online, they kept up to date on quilting shows through their friends' blogs, and they exchanged patterns with each other online. Online technologies had widened their communities and their access to quilting knowledge.

For our participants, discussions of mobile technologies and cellular telephones often slipped into a more general debate on all digital devices and telecommunication technologies. This "imprecision" is meaningful as it conveys how our participants experience the panoply of communications around them: cellular telephones are not discrete tools but part of a larger ecology of communication and telecommunication devices in their homes and in their lives. Our group of seniors may limit their cell phone use, but conversations indicated that their decisions were not made because they were digitally ignorant. They possessed a variety of communication options they would utilize when appropriate, including Skype and email. One woman recalled the days when she used to be afraid of the computer: "I guess I got my first computer when I was about 49 or 50. And uh, yeah, I was afraid to touch anything. I thought it was going to explode, actually. Now I'm a little past that" (Elizabeth, 64). The cell phone was positioned as but one digital device used by this cohort, all of whom were heavy users of the Internet, participating in blogs, chat rooms, and discussion groups, deploying a variety of tools to communicate with friends and family in faraway places. They spoke proudly of using their computers to perform their duties as mothers and grandmothers, including using Skype (for free!) to make customized curtains for a family in another city. Although their cell phone practices are restricted, the Lakeside Quilters are fully engaged with mediated forms of communication. This group's online sophistication meant that they did not need the cell phone for anything other than as a communication device for talking in very specific local contexts that minimized costs, but provided another option if and when needed.

Tactical Learning: Dealing With Ageism

One barrier to cell phone use is the overt ageism that is a part of our society, and which members of this group experienced when asking for information when buying a phone. Several recounted how they would carefully prepare a list of questions before going to buy a phone, or returning to a store for information only to have their list greeted with disdain by impatient salespeople. They had particularly harsh words for retail staff who showed "a little bit of disrespect because you can't pick up something as quickly as a ten year old, but it doesn't mean you want to be left behind" (Jean, 74).

These stories of disrespect, by more than one person at the table, brought forth a sharing of tactics to combat it. One tried-and-true mechanism used by more than one woman was to make the ageist assumptions work in their favor. In the words of Doris, eighty-four years of age, "I use my age to get what I want":

> I have found it convenient to use age, you know, if I go into a store, and I don't understand something, I say: "You know, I really don't understand that. I'm 84 years old, and I don't understand it." And it's amazing how it works, particularly with young men [laughter]. (Doris, 84)

Doris mobilized ageism and sexism to get the information that she needed, a tactic of what one might call ironic deference that uses one's position of subordination in a knowing manner to extract the needed information. This explicit, ironic use of ageism was often accompanied by the request "to speak slowly" and the deliberate use of stereotypes such as the "electronically challenged senior" (Carol, 75). These women performed "old age," reinforcing the expectations that the young have towards the old.

And here it is worth returning to de Certeau. Tactical appropriations may not involve systemic transformation. Instead they involve the ancient operational logic of disguise and survival. De Certeau considers this as a form of "poaching" or *la perruque*, a term that literally means to be costumed in a wig or disguise. It is through such performative means that one builds a political "poesis." De Certeau's articulations of the tactical maneuvers made by individuals, or groups of individuals, always returns us to their implication in a plurality of relations, a "series of organizational combinations" or entangled mesh of intertwining parts.

Yet while the specific context of the store may have demanded that they play the role of a feeble-minded old person, our participants had harsh words for instances of unintentional ageism they had experienced. Several times the problem of impatience in explaining, or over-explaining, or grabbing the device out of the user's hand were singled out. Here the question was often one of speed in the delivery of information and in giving instructions, a nod towards subtle shifts in cognitive abilities, but as well an admission of the pace of life in the digital age:

> I think it's the fastness of it though. You know, just even watching my son—who's in his mid-30s—you know, he flips through those channels, and the kids do it all the time too, they just go, you know, ch ch ch [remote control sounds], and I'm going "whoa, whoa, whoa, what's on?" You know, like, do do do do [imitating slowness]. And it's, it's fast...it's too fast. (Elizabeth, 64)

The quilters acknowledged that there were physical changes brought about by old age that made using such a small device cumbersome, including the little buttons and the screen size, yet these design issues were not front and centre of their concerns. Here it is worth noting that these are women who are dexterous, and whose hobby depends on their ability to keep arthritis at bay.

Yet while they noted changes in ability, they questioned whether the real difference between being young and being old was physical or psychological:

> Don't you think a lot of us are afraid? Like, I have a number of friends that call me and say, "Would you please do this," or "Find this on the computer," or "Do this on the computer." But their biggest fear is that they're going to do something wrong. And they find that their kids aren't very helpful: "Well mum, I'll just do it for you," and that doesn't help them at all. And I think...we get the feeling that we can't do it, and we can. (Ruth, 70)

They marveled at the way that the younger generation seemed to have no "fear" of using the phone, or any digital device for that matter. Several noted, and nodded in agreement, when one of them mentioned that they lacked confidence when it came to testing out and using the device.

Yet, they also made a point of stating that age was not solely to blame for troubles with complicated cell phone menus. Complicated menu structures and limited use was not just their problem. As one quilter quipped, "My daughter barely knows how to use her smartphone" (Carol, 75). Another added, when you get older, you are "a little tired" and sometimes this requires "more effort" (Elizabeth, 64). All were quick to point out that being slow was not the same as being stupid or incapable of learning.

One of the recurrent comments by the quilters was how the design of the phones, as multi-purpose, multifunctional devices, as well as the sheer number of phones on the market made learning these different functions difficult. As this exchange between participants reveals, it is the lack of consistency in the design that makes learning a challenge:

> I think it's the lack of consistency. I mean, originally, when we had a rotary telephone, all rotary phones were the same. You knew how to dial. If you get a Telus phone or a Fido or a Rogers phone, and even variations within those different places, they're all different. (Evelyn, 61)

> That's correct. I have experienced that. (Jean, 74)

> There's no consistency there. And I think that's because each manufacturer or whatever wants to have its own features that will sell the product. (Francis, 71)

What was helpful, at least according to one participant, was the realization that the cell phone was less like a regular phone, and more like a little computer, a digital technology the members of the group were comfortable using. This realization made her instantly more comfortable with testing it out.

These discussions of modes of learning—asking questions and getting information— underscore this group's relationship to print and oral culture. As one participant put it, "I think the difference in the generations is we were raised where we read the instructions, and then we went ahead and did it. Now they don't want instructions. There's no instructions, they just do it" (Francis, 71). They candidly repeated, several times, that they rely on manuals for information: "I have to get the manual, and read the manual," said Helen, 84, while others counted on family members. They are keen to learn, but on a need-to-know basis and they like a "go to" resource like a manual.

Yet what also became evident was that the sessions were seen as an environment where group learning could take place. Like many other sessions we have held, more advanced users offered tips and advice on how to negotiate with companies to get a better deal as well as how to use particular features of the phone that they found useful, such as the alarm function. Here we return to Virpi Oksman's research on age differences between users in the Finnish context. As her research acknowledges, more experienced users often play an important role in teaching and sharing tips about mobile communication with their peers. As she noted:

> The age group of seniors thus often has its own innovators who have an important social role in their circle of friends in the spread of new mobile devices and their uses. They are often mediators of information connected with mobile phones and computers: advice in the purchase of devices and the choice of operator. (2006, p. 15)

Seniors have their own pool of experts who may provide guidance to others.

Like our Lakeside Quilters, Oksman's Finnish seniors are concerned by the "blurring" of the private and public spheres of communication, yet, rather than characterizing their reflections and critique as fear or hesitation, she uses the term, *caution* (2006, p. 14). Understanding the imposition of these measures as cautious, tactical maneuvers underscores the political and poetical dimension of the quilters' mobile media practices within their specific local contexts that takes differences into account.

Conclusion

We all age, but how we experience age can differ radically in different national or local contexts where subjects live under very different conditions of power and privilege. When it comes to digital technology the quilters we have spoken with are not naive non-adopters, victims, aloof, or blindly resistant. Many are, to echo Oksman's (2006) words, "cautious." But they are also critical and skeptical of the promise of technology. They have, after all, seen it all before, having experienced the introduction of a variety of technologies throughout their lives including radio, television, and the Internet.

The Lakeside Quilters, like the scores of seniors we have now spoken to across Canada, are a part of what is happening with the world around them. Even if they do not own every piece of new technology, deploy every function on a mobile device, or use their cell phones to maintain constant access to family and friends, these users are well aware of new trends in telecommunications. Within this group of women, there is a willingness to adapt and to learn. But there is less of a compulsion to keep up with technology for its own sake. Their rules for engagement reflect their economic circumstances as people who are no longer part of the paid workforce, who have a particular set of social values they cherish, and who deliberate on what communication devices best suit their needs. They may not use multiple functions on one device, but instead use the multiple devices that they have bought, and learned, including the landline phone, the computer, and, when needed, the mobile telephone.

There are some aspects that make the Lakeside quilters unique: they are a group of women who have had paid employment in their lives and who now have a shared passion that puts them into contact with digital technologies. They are located in a rural area that is, nevertheless, close to a metropolitan hub, which guarantees a modicum of access to cellular services. However, in their observations we have heard echoes of shared sentiments with other discussions held with seniors across the country. To return to the comments of de Certeau (1984), there are a range of tactical maneuvers that are used by this cohort to deal with living in a rural area where driving is exceedingly important, living on a fixed income, and addressing their encounters with systemic ageism. These senior women use their cell phones tactically, with forethought and a refreshing wariness to digital hype and the latest set of promises that their worlds are about to change because of the latest 3G technology for "keeping in touch." In all but a few cases, digital technologies and new modes of communication are not being rejected outright. When fear was expressed it was not expressed as a lack of confidence, nor as the result of the surprise of an unexpected billing horror. Like many of the other seniors we have spoken with, they tactically restrict their use of communications technologies as they have entered into their post-retirement lives for a variety of reasons. Of these, the cost and restrictiveness of the service contracts is significant. But their decision to restrict their usage also reflects a different set of values regarding money, time, and the need to be in constant communication. Restricting communication, managing their cell phones, developing "rules" for its careful integration into their lives, and guarding private time enhance their sense of agency.

This emphasis on the "practices" of everyday life and tactical media use shares insights with researchers, such as Sally Wyatt (2005). In her reflections on the practices of users and non-users in Internet studies, she questions the presupposition that using is always better than not using; and usage is confused with agency. As she wisely states:

> The Internet "user" should be conceptualized along a continuum, with degrees and forms of participation that can change. Different modalities of use should be understood in terms of different types of users, but also in relation to different temporal and social trajectories.... Internet use encompasses not only different types of use, but also the possibility of reversals and changes of direction in the individual and collective patterns of use. In addition to the usual demographic variables, details about the frequency and nature of use help to construct a fuller image of the multiplicity of uses and users of the Internet. (2005, p. 77)

In the present context of low-cost landlines, the widespread use of computers and broadband services, and the high costs and confusing service contracts, as well as other variables, make cell phones seem but one option for our seniors to communicate with family, friends, and loved ones. Perhaps the most important lesson and point of reflection from our conversations with the quilters is this: the statistical data that portray those fifty-five and over as inert, inactive, disinterested, and a homogenous group of users with technological "limitations" is simply wrong. Such a presupposition about their stubbornly reluctant engagement with the present mobile moment belies their actual practices of deliberate, tactical restrictions.

Acknowledgment

We would like to thank Ana Rita Morais for her research assistance, and the SSHRC for funding this work.

Notes

1. This is not a phenomenon particular to mobile media studies, where we see a disproportionate neglect of elderly audiences from within media, communications, and cultural studies, with a few notable exceptions, including texts by Blaikie, 1999; Counts & Counts, 2001; Featherstone & Wernick, 1995; Riggs, 1998; Walker, 2007; and Woodward, 1999.

2. See Statistics Canada's (2007) *A Portrait of Seniors in Canada,* http://www.statcan.gc.ca/ads-annonces/89-519-x/index-eng.htm

3. "By the end of 2007, the world had: 3.3 billion mobile cellular subscribers, 1.3 billion fixed telephone lines, 1.5 billion Internet users, and 336 million broadband subscribers" (International Telecommunication Union's (ITU), "World Telecommunication/ICT Indicators Database." Retrieved from http://www.itu.int/ITU-D/ict/statistics/ict/index.html

References

Blaikie, A. (1999). *Ageing and popular culture.* Cambridge, UK: Cambridge University Press.

Brants, K., & Frissen, V. (2005). Inclusion and exclusion in the information society. In R. Silverstone (Ed.), *Media, technology and everyday life in Europe: From information to communication* (pp. 21–32). Aldershot, UK: Ashgate.

Canadian Association of Social Workers. (2004). *Women's income and poverty in Canada revisited.* Retrieved from http://www.crvawc.ca/documents/Women%27s%20Income%20and%20Poverty%20Revisited%20-%20Executive%20Summary%20-%20English%20Version%20.pdf

Canadian Wireless Telecommunications Association (CWTA). (2008). *Harris/Decima 2008 wireless attitudes study.* Retrieved from http://www.cwta.ca/CWTASite/english/ industryfacts.html

Caron, A. H., & Caronia, L. (2007). *Moving cultures: Mobile communication in everyday life.* Montréal, Canada: McGill-Queen's University Press.

Castells, M., Fernández-Ardèvol, M., Linchuan Qiu, J., & Sey, A. (2006). *Mobile communication and society: A global perspective.* Cambridge, MA: MIT Press.

Charness, N., & Boot, W. R. (2009). Aging and information technology use: Potential and barriers. *Current Directions in Psychological Science, 18*(5), 253–258.

Conci, M., Pianese, F., & Zancarano, M. (2009). Useful, social and enjoyable: Mobile phone adoption by older people, *Lecture Notes in Computer Science (LNCS), 5726,* 63–76.

Conference Board of Canada. (2009). *Elderly poverty.* Retrieved from: http://www.conferenceboard.ca/hcp/details/society/elderly-poverty.aspx

Counts, D. A., & Counts, D. R. (2001). *Over the next hill: An ethnography of RVing seniors in North America* (2nd ed.). Peterborough, Canada: Broadview Press.

Cresswell, T. (2006). *On the move: Mobility in the modern Western world.* New York, NY: Routledge.

de Certeau, M. (1984). *The practice of everyday life.* Berkeley, CA: University of California Press.

Featherstone, M., & Wernick, A. (1995). *Images of ageing: Cultural representations of later life.* New York, NY: Routledge.

Goggin, G. (2006). *Cell phone culture: Mobile technology in everyday life.* New York, NY: Routledge.

Gow, G. A., & Smith, R. K. (2006). *Mobile and wireless communications: An introduction.* London, UK: Open University Press.

Gregg, M. (2004). A mundane voice. *Cultural Studies, 18*(2-3), 363-383.

Haddon, L. (2005). Empirical studies using the domestication framework. In T. Berker, M. Hartmann, Y. Punie, & K. Ward (Eds.), *Domestication of media and technology* (pp. 103-122). Maidenhead, UK: Open University Press.

Horst, H. A., & Miller, D. (2006). *The cell phone: An anthropology of communication.* Oxford, UK: Berg.

International Telecommunication Union (ITU). (2007). World telecommunication/ICT indicators database. Retrieved from http://www.itu.int/ITU-D/ict/statistics/ict/index.html

Ito, M., Okabe, D., & Matsuda, M. (2005). *Personal, portable, pedestrian: Mobile phones in Japanese life.* Cambridge, MA: MIT Press.

Katz, J., & Aakhus, M. (Eds.). (2002). *Perpetual contact: Mobile communication, private talk, public performance.* Cambridge, UK: Cambridge University Press.

Kurniawan, S. (2008). Older people and mobile phones: A multi-method investigation. *International Journal of Human-Computer Studies, 66*(12), 889-901.

Lenhart, A. (2010). Cell phones and American adults: They make just as many calls, but text less often than teens. Pew Internet & American Life Project report. Retrieved from http://www.pewinternet.org/~/media//Files/Reports/2010/PIP_Adults_Cellphones_Report 2010.pdf (no longer accessible)

Ling, R. (2004). *The mobile connection: The cell phone's impact on society.* San Francisco, CA: Elsevier/Morgan Kaufmann.

Ling, R. (2008). Should we be concerned that the elderly don't text? *Information Society, 24*(5), 334-341.

Middleton, C. A., & Cukier, W. (2006). Is mobile email functional or dysfunctional? Two perspectives on mobile email usage. *European Journal of Information Systems, 15*(3), 252-260.

Mohd, H. N., Hazrina, H., & Nazean, J. (2008). The use of mobile phones by elderly: A study in Malaysia perspectives. *Journal of Social Sciences, 4*(2), 123-127.

Morris, B. (2004). What we talk about when we talk about 'walking in the city.' *Cultural Studies, 18*(5), 675-697.

O'Brien, M. (2009). The tactics of mobile phone use in the school-based practices of young people. *Anthropology in Action. 16*(1), 30-40.

Oksman, V. (2006). Young people and seniors in Finnish 'mobile information society.' *Journal of Interactive Media in Education, 2*, 1-21.

Pierson, J., Mante-Meijer, E., Loos, E., & Sapio, B. (2008). *Innovating for and by users.* Brussels, Belgium: European Science Foundation.

Potter, J., & Wetherell, M. (1987). *Discourse and social psychology: Beyond attitudes and behaviour.* London, UK: Sage.

Riggs, K. E. (1998). *Mature audiences: Television in the lives of elders.* Piscataway, NJ: Rutgers University Press.

Riggs, K. E. (2002, June). *The new, new deal.* Paper presented at the Informing Science:

InSITE—Where Parallels Intersect conference, Cork, Ireland.

Riggs, K. E. (2004). *Granny @ work: Aging and new technology on the job in America*. New York, NY: Routledge.

Statistics Canada. (2010). *Average income by family types*. Retrieved from http://www40.statcan.ca/101/cst01/famil21a-eng.htm (no longer accessible)

Statistics Canada. (2007). *A portrait of seniors in Canada*. Retrieved from http://www.statcan.gc.ca/ads-annonces/89-519-x/index-eng.htm

Urry, J. (2007). *Mobilities*. Cambridge, UK: Polity Press.

Wajcman, J., Bittman, M., & Brown, J. (2008). Intimate connections: The impact of the mobile phone on work/life boundaries. In G. Goggin & L. Hjorth (Eds.), *Mobile technologies: From telecommunications to media* (pp. 9–22). New York, NY: Routledge.

Walker, A. (2007). Why involve older people in research? *Age and Ageing, 36*(5), 481–483.

Wilken, R. (2005). From stabilitas loci to mobilitas loci: Networked mobility and the transformation of place. Retrieved from http://six.fibreculturejournal.org/fcj-036-from-stabilitas-loci-to-mobilitas-loci-networked-mobility-and-the-transformation-of-place/

Wireless North. (2010, August 27). It's 2010 and Canadians pay the highest phone bills in the world. Retrieved from http://wirelessnorth.ca/

Wong, C. Y., Thwaites, H., & Khong, C. W. (2008). 'Oh! My battery was drained because I forgot to press the end call button.' In *Proceedings of the 21ˢᵗ International Symposium on Human Factors in Telecommunication: User Experience of ICTs, Kuala Lumpur* (pp. 31–38). New York, NY: Prentice Hall.

Woodward, K. (Ed.). (1999). *Figuring age: Women, bodies, generations*. Bloomington, IN: Indiana University Press.

Wyatt, S. (2005). Non-users also matter: The construction of users and non-users of the Internet. In N. Oudshoorn & T. Pinch (Eds.), *How users matter: The co-construction of users and technology* (pp. 67–80). Cambridge, MA: MIT Press.

10

Haunting Technologies: Performing Memories of Place Through Effervescent Mobilities

Phillip Vannini and Rhys Evans

In the spring of 2010, we—Phillip Vannini and Rhys Evans—met in person for the first time in Victoria, at the Cultures of Movement conference. Well before the conference, though, we had gotten to know each other a bit by email. We had talked about the conference and had begun to share a little bit about our personal lives. Phillip had learned that Rhys was not a Scotsman, though he was living in Scotland at the time. Rhys had called home the mainland of British Columbia and then Vancouver Island—having lived in Victoria and Nanaimo—as well as Gabriola Island. At the time Phillip was living on Vancouver Island and was planning on moving to Gabriola Island with his family. When Phillip told Rhys about his research—a four-year, mobile ethnography on the role played by BC Ferries[1] in the lives of island and coastal residents[2]—Rhys immediately began to share stories and memories about catching the ferry, both as an islander and as a truck driver. Phillip reciprocated with some stories and memories of his own. The conversation continued at the conference and after.

It is extremely common for islanders of the world—both current and former—to shoot the breeze about island life and the technologies of mobility, such as ferry boats, that make both island life and the subjectivity of islanders unique. Through these conversations islanders perform their identity and contribute to the constitution of the roles which islands and ferries play in their day-to-day lives. These conversations punctuate the days, weeks, and seasons of islanders' lives. Indeed chatting about the ferries felt so natural for Rhys and Phillip that they thought that rendering their collaboration on this chapter in any way other than through dialogue would be an act of ethnographic misrepresentation. The following is thus a polished, autoethnographic rendition of that conversation, focused on the spectral geographies (see Holloway & Kneale, 2008; Maddern & Adey, 2008) of mobility technologies.

Rhys: Do you remember the first time you caught the ferry to Nanaimo?

Phillip: I do. It's a pretty vivid memory, actually. It was a special trip; I was travelling from eastern Washington State to Nanaimo to come and visit April, then my girlfriend. It was spring break, 2001. I distinctively remember looking at Nanaimo's lights in the darkness of the evening, and trying to imagine if that would be the first of a long series of trips.

Rhys: Did you have a feeling it wasn't just a spring break fling?

Phillip: Yeah, definitely. I was two or three years away from finishing up my PhD, so I knew that if our relationship was to last I'd have to catch a few ferries before we could find a way to move closer to each other. That image of Nanaimo from near the Entrance Island lighthouse, you know the one off of Gabriola, is completely burned in my consciousness. I was at the front of the ferry, either the *Queen of Alberni* or the *Queen of New Westminster*—I can't remember—just looking out the windows, just wondering where the hell I was going. And then one of the crew came by and put up black tarps on the windows, and I couldn't see anything anymore. And that was good because it stopped me from thinking too much. Though, it took me two years to figure out why they cover up the windows at night.

Rhys: To minimize the glare from the lounge, so the bridge can see better.

Phillip: Right. And that's the thing. Even though I didn't know what I was doing there, or where I was going, there was this odd feeling of being part of a ritual greater than my own fledgling commute—a ritual that I could feel, yet not quite yet understand. It was like being in a bit of a liminal space, you know, going with the flow. I had this feeling that I was in some kind of a portal and that after that moment, after that trip across the water, life would never be the same. Have you ever gotten a feeling like that?

Rhys: You know, for me, the ferries weren't a place of anticipation so much as a place of certainty. In a journey filled with little unknowns—will I make the five o'clock sailing, will the car make it to Gold River over the dirt logging roads?—the ferry was a constant, a safe place of comfort. It was familiar, predictable. Almost like a home.

Phillip: I so know what you mean. That feeling of rush in trying to catch the ferry, and then that feeling of calm once you've made it....

Rhys: Yeah. And what I always used to notice was the rhythm of a ferry trip. It always had three very separate speeds, almost like movements in a symphony. First was the rush to get to the ferry on time: always leaving at the last minute, cursing traffic delays, racing over the Upper Levels Highway to Horseshoe Bay. There was always the constant relief when you hit the wide open stretches above West Vancouver and the traffic began to speed up, sweeping round those wide curves so high up the mountainside with those superlative views out over Georgia Strait. It was the first time you noticed the weather, looking out towards the Island. Then everything stopped. And you're in the queue. "Will we make the sailing? Will they let us on?" And if not, perhaps you got out of your car with all the other folks and idly leaned against it in the sun. After that, a momentary flurry of activity as you finally get loaded, driving into the dark, cave-like maw of the ferry's boarding mouth out of the brilliant sunshine. And then you stop again. And that's it, till you get to the other side, apart maybe from the line-up for the cafeteria. Finally there was the rhythm of the journey on the other side. Slower than the rush to catch the boat, but back engaged in negotiating your own journey yourself, under your own control again. So, three movements—starting with *allegro*, then a kind of *adagio*, and a final movement of *legato*. The symphony of a ferry trip to the Island!

Phillip: Absolutely, I've noticed that too. Hey, do you remember the first ferry trip you ever took?

Rhys: You know, I don't. Growing up here, ferries were always part of my life. I suppose it was family holidays in the summer. We used to spend a week in a cabin somewhere. Once at St. Mary's Lake on Saltspring Island; once at Miracle Beach up between Courtenay and Campbell River on Vancouver Island. And when I was a teenager, taking the ferries was part of my wanderlust—many trips to the West Coast, hitchhiking to Long Beach. And that was long before this Tsawwassen–Duke Point run existed. They were the old ferries then. Smaller. Much more civilized, with proper restaurants and staterooms, even. Not that we, or anyone we knew, ever took a stateroom! That was one of the great things about the BC Ferries. They were the great leveler. Before all these little airplane services, the ferries were the only way to get to the big island. And there was no "Preferred Boarding." It didn't matter who you were, rich or poor, you took the ferry. And it took just as long for everyone. Whether you were driving an old '52 Chevy pickup covered in rust or a brand new Cadillac, you drove up to the ticket booth, paid your fare, and waited. Then you drove on the ferry and went upstairs where you sat with everyone else and the trip took as long as it did. Your kids played on the same carpet. You ate the same food. And you waited as long as everyone else did until it got in at the other side. The great leveler.

Phillip: That is one of the most common memories about the ferries I've heard in four years of fieldwork. I remember this man from northern Vancouver Island tell me exactly the same thing, even calling the ferries a "leveler" himself. Back then he and his family lived in a shed in the bushes. They'd get on the ferry and head to the washrooms to clean themselves up, using the soap, the sinks, and the running water to look presentable before they'd go to the lounge. Not that looking presentable mattered that much—as there were a lot of people who were just coming out of the bush—but it was a very remarkable, special occasion, something that punctuated an extraordinary experience. Whether it was visiting family on the mainland, or whatever occasion, the ferries stick out in many people's memories as being a common space, a shared place where memories of a longer journey begin, you know...?

Rhys: Like in the expression "we're all on the same boat..."

Phillip: Exactly! So many people tell me they remember that everyone had to begin their journey the same way, and everyone was treated as equal.

Rhys: That's for the big ferries to Vancouver, though. Do you think it was the same for the smaller routes?

Phillip: Yes and no. As you know, the smaller islands have a different relationship with their ferries. They're smaller communities, with smaller boats. Whereas getting to the mainland is a common affair that brings together all kinds of people, getting to and from a smaller island is a more selective process.

Rhys: What do you mean by *selective*?

Phillip: A man from Denman Island explained this really well to me. He told me a story that many people on the island remember. It's somewhat of a myth. It's unclear what actually happened because memory is selective and in retelling it over the years the story has evolved, but it's told like this.

There was a minister on Denman who had a habit of showing up for the ferry back to the island at the last minute. People had gotten used to that, including the captain. He was well liked and respected, and the captains were kind enough to even wait for him for a few minutes before leaving Buckley Bay. Sometimes he'd arrive early, park the car, and go visit with people at the store near the terminal. One day, as he did that, the ferry started loading. It was a busy day and the ferry loaded quickly. Indeed so quickly that soon enough there was only one spot left on the car deck. As he rushed back towards his car, another car—a nice shiny new convertible—arrived at the terminal. The convertible drove around the priest's car, and took the last spot onboard, leaving the poor fellow behind, having to wait for the next sailing. The islanders on deck saw this happen. As it turns out the passengers of the convertible—people from the city—got out of their car and walked towards the front of the ferry, to enjoy the view and the fresh air from there. By the time they got back to their car in the back of the boat, their wheels were gone, and their car was resting on wooden stilts. Nobody had seen anything, of course, and the wheels were nowhere to be found. It was like a ghost had done that.

Rhys: That's a pretty good story. I wonder how much of it is true.

Phillip: That's the thing—no one really knows. But in one way it doesn't really matter, right? What matters, as this Denman Islander told me, is that almost everyone on Denman knows a version of this story. By telling it to one another, as he said to me, they're conveying the idea that ferries are not only levelers, but they are also selective mechanisms, you know? They're part of rituals that are inclusive for an in-group, but they are also exclusive for an out-group, and that factor of exclusivity is really sharp on small islands.

Rhys: So they are *levelers* in all sorts of ways—they bring stuck-up outsiders down to "our" level, and they bring us together with a shared sense of equality, right?

Phillip: Yep, absolutely! I gotta ask you, Rhys, as a trucker did you guys have some sort of members-only club on the ferry, with your own myths, your own rituals, and stories?

Rhys: Well, just like there are "regulars" from the community on the small island runs, on the big ferries there are communities of regulars, too, only they are the people who use the ferry as part of their job. So yes, there certainly was a truck driver community when I was going back and forth. The ferries are a bit like a small city and the regulars form communities within it.

I first started taking trucks on the ferry when I was hauling produce from California to BC. Sometimes I would have a load to take to Nanaimo or Victoria and I would go straight there from Los Angeles. In those days I would nod my respects to other truck drivers while on the boat, but that was about it for me. But in the early 1980s I started driving line-haul for a courier company, and that took me over to Vancouver every day, from Nanaimo. I would start at 1:00 p.m., catch the 3:00 p.m. ferry, move a couple of trailers round Vancouver and catch the last ferry back to Nanaimo. T'was a great job—I used to get 1½ hours overtime every shift and spent nearly four hours of it sitting on the ferry or in the loading parks.

One of the cool things about taking trucks on the ferry is that, after motorcycles, you were first on and first off. So you would park up the truck and head up to the passenger deck before other people were getting out of their cars. You would go right away to the cafeteria where there was this informal place for truckers to sit. It was located right under the sign which said, "No Smoking, No Gambling." And that was significant, because, located under a seat cushion right underneath the sign was always a deck of cards. The drivers would play poker—y'know "thirty one"?—with a dollar in the pot for each player. The one with the last hand would take the pot.

Phillip: Right under the "No Gambling" sign...

Rhys: Yes! And depending on how many people were playing, you could walk away with twelve or fifteen dollars—enough to pay for your food on the boats and a couple of bucks for beer on top. They were great games, with lots of joshing and bluffing, and of course, being engaged in this particular, slightly naughty activity signaled membership in a special community. The ship's officers used to always sit right next to us and the games were always covertly acknowledged and accepted. They would laugh at our jokes and sometimes take an interest in a game.

Phillip: Did this happen all the time?

Rhys: Hmm, it usually only happened on the late sailings. In the daytime I would usually sit outside on the deck, or even lie in the sun. But after dark, everyone gathered round to play cards. One night we were playing and I was winning. There was $14 in the pot and we were down to the last two hands, and I was still in it. The ferry actually docked while we were playing the last hand—all the announcements had been made and all the passengers were already down waiting in their cars—and we played on. And I won! We all knew they had to wait for us because our trucks were at the front of the lines down below, but we were so excited! And, the ship's officers indulged us, knowing we were hustling to finish. So when done, I scooped my winnings up and we all ran down the steps, jumped into our trucks, fired them up and drove off, laughing at the mystified looks on the faces of the car drivers who had been sitting, waiting patiently to be allowed to unload!

Phillip: Oh my god, that's awesome. I would have killed to be part of any of those games as an ethnographer!

Rhys: Like the story of the minister on Denman Island, our *belonging* was signaled by the tolerance of the captain and crew. Just as they waited for the minister to arrive, so they waited for us to depart, knowing that we, as insiders, recognized the privilege we were being accorded.

You're living on Gabriola Island now, Phillip. Now *you* are a regular. Has this changed your own experience of the ferries, being a real insider?

Phillip: I catch the ferry about twice a week, every week. The key difference is that for the past four years catching the ferry was a form of work. I was a sensory sponge—nothing could possibly escape my field journal. Now, I tell myself to relax and just view the ferry as a means to my work day, rather than the subject of my work day. So, for example, when I catch the 6:40 a.m. to Nanaimo I often fire up my laptop and do work in the car. At times I do the same at night, when I catch the 6:10 p.m. back home from Victoria. At other times, I take a quick catnap. I am very insulated in my car, very alienated from the people and the place around me.

It's completely different on weekends when together with April and the kids we go to Vancouver Island for shopping, errands, or leisure. Especially because Autumn, my five-year-old daughter, is with us. She's a social butterfly—she knows half the island, I swear. So, she wants to get out of the car and go visit with people in the lounges, on the deck if it's a good day, or in people's cars if any of her little friends are driving to Nanaimo with their moms and dads. So I kind of follow her around and end up visiting with her friends' parents.

Rhys: So catching the ferry has become more routine...

Phillip: Definitely, but that doesn't mean the spark of the ritual has gone. If anything it's gotten stronger. I remember a summer morning, just last year, when my stepson Jake and I grabbed a football out of the trunk and started playing catch on the car deck. There weren't that many cars, so we had space for 20-to-30-yard passes, and for running catches too. The ball almost went overboard a couple of times. It was a blast. And I've seen other people doing the same, you know. If not with a football, with a soccer ball, or a hackey sack. For me that memory stands out as being such a vivid island life moment. You couldn't do that if you were stuck in your car on the parkway.

Rhys: And indeed, you wouldn't probably do it the first time you rode on a ferry. Appropriating the space for our own vernacular uses is one of the ways we once again demonstrate our belonging, our status as *regulars*.

Those are good memories for the kids, too. That's how you grow up feeling an islander. I call it "running around like they own the place" and mean it in a profound sense.

Phillip: Yeah, and then you think about those kids whose expecting mothers depend on the ferries on a middle-of-the-night emergency rush to the hospital! Think about the stories they grow up with....

Rhys: Have you met any of those?

Phillip: About four or five, between kids and mothers. The kids don't remember much...

Rhys: No shit!

Phillip: Hehehe! But the mothers' stories are unforgettable: feeling the labor is beginning and realizing the last regularly scheduled ferry of the night has left, panicking, then getting on the phone to call the ferries, waiting till a captain and crew have been located, then rallying to the ferry terminal and realizing the ferry is waiting for you and only you, and half the crew are in their pajamas...

Rhys: That's just too much!

Phillip: I think that the best example of the power of memories for most people is when you are returning home from a long trip away, far away. So many people have told me about these blissful moments, when you get back to the ferry terminal and start to smell home, you start to see it, to feel it in your blood. And then you get on the ferry, and the ritual of travelling home just overpowers you. Someone from Sointula shared a pretty special memory with me a while ago. They had just come back from a long trip. They were tired, homesick, just couldn't wait to get back to their island. But all the fatigue and the nostalgia began to evaporate as they arrived at Port McNeill and lined up for the ferry. It felt as if they were home already. Then they drove on the ferry. It was a warm summer night, with the stars lighting up the sky. Somebody got out of their car carrying a guitar and started playing, "Don't worry, be happy." Within seconds almost everyone on the car deck got out of their vehicle and started singing along. And then they kept on playing. It was like a moment out of time, I was told, but one of those moments that really burn a place deep in your heart.

Rhys: Wow.

Phillip: What's it like for you now when you get back to the island? You've been gone to Europe for a while now—do you feel any nostalgia when you get on the ferry?

Rhys: Oh ya. Especially when it is the older ones—the *Alberni* and *Coquitlam*—the ones I used to ride on. The corners are haunted by all the things that happened in them. In fact, it's a bit strange now, because of some of the changes, like the no smoking on the outside deck except in the one place. On the way over to Vancouver I used to take a towel and shorts and sit out getting a tan, smoking, and drinking coffee. I would pick the "sweet spots," like against the starboard side between two life-jacket lockers, where the wind would be low and on the sunny side. I tell you, when I had that Loomis job I was tanned by the end of May! And every day there would be people—different individuals, of course, but somehow still a set of types. There were the older got-it-together hippies coming over to Vancouver or going back to the Island who would come up and smoke joints. In the old days, the deckhands didn't care if you toked up on the deck and it was quite a social place. Now, of course, they'd come and tell you to put your "cigarette" out. I watched them do it last time I took the ferry.

Phillip: Right. Though, still today, if you talk about the smellscape of a BC Ferry, you have to talk about that particular smell first and foremost...

Rhys: Too true. But anyway, the thing about the ferry is its constancy. Despite the changes, so much is the same. Like the look: I had a good friend, Jim MacKenzie from Mayne Island, who is an artist, and he created all these super-realist paintings of archetypal ferry bits, the railings with the sea and mountains behind; the colours—brilliant white and blue with the greys and greens of the rest of the environment behind. And even now, when I get on the ferry, I see his paintings.

There are other ghosts, too. Like the people. There are new "types" but still all the old ones. There's the First Nation kids running around on deck, the older hippies with the wind in their hair, the middle class families with the young kids. There's the loggers and other working men in their "Port Alberni dinner jackets." You know those?

Phillip: Oh my god, I can't believe you just said that! I heard of "Cedar dinner jackets" and "Harewood dinner jackets," but I didn't know the couture line had made it to Port Alberni too!

Rhys: Yep, it's the same. Those red or green, black-checked heavy shirts that loggers wear. Well, in the logging communities up and down the island a newish one, freshly laundered, was considered good enough to go out in to the most expensive restaurant in town. And they wear the ball caps that say things like "Vancouver Island Helicopter Logging," or "Reg Dorman Trucking." Serious signifiers not only of their jobs but of their way of life and their identities. Of course there are new types, too—the young media geeks going home for mom's home cooking, the gap-year traveler girls with tanned legs, cargo shorts, and backpacks. But even they are just new versions of an old story.

Phillip: And don't forget the first-time tourists. The Americans trying to come to terms with the very existence of islands they had never heard of before...the Europeans taking pictures of everything moving in the water while looking for whales...

Rhys: Yep. You would think that it was people who are the ghosts, but it is also the stories, or perhaps the actual events behind them. And indeed, every story has a piece of the ferry involved in it—the car deck, for example. My ex-wife worked as a cleaner on the Nanaimo–Horseshoe Bay run when I was driving back and forth every day, and sometimes we'd be on the same boat. So I would sit in the Crew Mess with her on coffee break. And every single day a deck hand would come in after doing their rounds and tell us about somebody making out in their vehicle on the car deck. "Over in the forward car deck, port side. If the van is rockin', don't go knockin'!" they'd say with a laugh. There are stories from the stairs, escalators, and elevators....

Phillip: And the ghost cars, those that are never found after the ferry docks. I've had so many people tell me that they once forgot they drove on, but they walked off and forgot their car onboard....

Rhys: And from the galley, the cafeterias, and forward lounge. And always stories from the washrooms.

Phillip: Oh yeah, the washrooms. The ghosts of early morning commuters, who after retiring still show up in the washroom to shave their beard and brush their teeth....

Rhys: Then there are the stories about the machinery down in the bowels of the beast. I was on a ferry one day coming into Nanaimo and it didn't go into reverse gear to slow down as we approached the ramp. We all just sat there incredulously as the ship came steaming in towards the dock unable to believe the master wouldn't slip it into reverse and stop the slow motion crash that eventually happened. Turned out the gear controls had broken. Did hundreds of thousands of dollars damage to the dock, which was out of commission for ages. I'll never forget the feeling of inevitability as we rushed towards the dock, and the huge bang as we bounced off it.

Phillip: Every single time I drive on the Vancouver ferry from the lower ramp, at Departure Bay in Nanaimo, I think of the day—back in '92—when that van from Alberta plunged into the water as the *Queen of New Westminster* moved prematurely from the dock. Remember that? Three people died. There can hardly be a worse death. I can feel their screams suffocated by the cold water piercing my ears, reverberating through my body, every time.

Rhys: Sometimes the boats seem almost animate. They are waterborne creatures with lives and characters of their own, after all. The big ferries are designed to flex—nothing that big can really be rigid or it will snap. If you sit in the very middle of the boat—you know these plexiglass panels on slides above the big windows? One day we were going across from Nanaimo and it was really rough. So rough that in the end we had to go around the back of Gambier Island to get to Horseshoe Bay, as the Williwaws came howling down from Howe Sound. I was sitting there and the boat was shuddering and shaking from the huge swells. They were coming in two directions, with one set of waves coming down the trough of another set at a 90 degree angle. And somewhere, just before we got to Bowen Island, the ferry gave an almighty shudder and the window I was sitting beside shattered! It just broke into a thousand shards of safety glass with an almighty crack! And apparently it happens fairly regularly, which is why they have those plexiglass panels ready to pull down when the glass goes. I always think of that when I see the panels.

Phillip: Some people will swear that boats have a life of their own. I've written about the *Queen of the North*, before and after her sinking (see Vannini, 2008), but there is another ferry that by all accounts had an uncanny life of her own: the *Queen of Victoria*. She had more accidents than most other BC ferries combined, but it wasn't just a faulty design issue. Crew who worked on her will swear she was haunted from the first day.

Rhys: Yeah, that big accident wasn't her fault.

Phillip: I've actually interviewed someone who witnessed the accident unfold in its entirety. This person was a teenager then, in 1970. It was an August morning, a very foggy one. My informant was on Mayne Island, near the lighthouse, hanging out with friends. They were looking at the water, at the *Victoria* going by. When it's foggy the ferries blow their horn every few meters, and on Active Pass the echo keeps bouncing against the walls of the mountains. It's very spooky in the fog because you hear the sound but sometimes you don't know where it's coming from. Then all of a sudden they saw a Soviet freighter appear out of nowhere. Within seconds it crashed onto the side of the *Victoria*, slicing right through her, making these gut-wrenching, screeching metal sounds, as if the boats were screaming. It's a miracle that only three people died. And then a few years later, in an incredible twist of fate, in the very same spot in Active Pass the engine of the *Victoria* caught fire.

Rhys: Man...

Phillip: What do you make of all this, Rhys? I mean, the memories, the stories... There is something pretty powerful about them, you have to admit. Speaking in basic sociological or anthropological terms it's pretty easy to say that in sharing these narratives we are performing a collective memory—a sense of time and place—but I don't know if I want to explain it just like that. It feels like I'd be explaining it all *away*. There is something mystic about these performances, something sort of evanescent which keeps coming back over the years....

Rhys: I think it's important to remember that all of these stories are *hauntings*, which emerge from people living their lives, individually and collectively. They are contingent in that they both structure people's experience and are structured by that very experience. We are interested in them in this case because we are interested in the ways that mobility structures, and is structured by, people's encounters with the "socio-technical assemblages" in which mobility is expressed. These hauntings are *emergent* in the sense that they occur in places, places where people, technology, and mobility intersect. And that is what I mean when I use the word *effervescent*. It is like a bottle of pop. First there is the liquid. Then you add the gas. But it takes an action, popping the lid off, to make the bubbles come. Here we have the place, the machines, and the people, but the meaning only begins to come up when something *happens* within and between all of these.

These contingent emergences are "structures" in the sense that, as you found in your fieldwork and as I know from my own experience, they are patterns that most of us agree we can see—whether ordinary ferry passengers or anthropologists! They bubble up out of the intersections of people's lives and livelihoods with the machines in which they move between places. But there is nothing given about them. They take the shape and carry the meaning that they do out of the particulars of all the here's-and-now's which constitute them. And in fact, it is exactly that particular-ness which makes them *BC* Ferries, and which makes the identities enacted upon them BC Coast identities.

Phillip: What theoretical angle do you take on this, then?

Rhys: Well, according to non-representational theory if we are to understand place as a truly *lived space*, then we have to go beyond words and representations to ways of understanding *feelings, affect,* and *embodiment.* Thrift (1999) said that places are *haunted* by these things and that these hauntings constitute what places are and become, what they mean to us, and indeed, who we are in them. And these ferry stories are a pretty good example of this. Thrift said that place is haunted by "the world of *things*" (1999, p. 312), and by three "full-bodied competences"—"emotion," "memory," and "language" (p. 314). We have already been talking about the "things"—the mechanical components of our ferry experiences. And we have pointed to the emotions—the feelings of coming home, going away, and all the other ways in which we respond to the particulars of our journeying. Further, Thrift said that "places are passings that haunt us" (p. 310). And here in this account, the BC Ferries are indeed "places" and we both "haunt" them as our past experiences imprint our current interactions with them, and they "haunt" us in the same way. Our stories are memories, reconstituted and risen with the yeast of individual experiences. And, by telling these stories, we are using language in the way Thrift meant when he said that language is "*performative,* a virtual structure achieved through use, not a potential structure actualized by use" (p. 315). That is, *we* create the realities the stories tell, and the stories create the *realities* we experience. So, in a funny sense, we create them and they create us. Phillip, you used the word *mystic.* What do you mean by that, and do you think these hauntings might constitute the specialness you are signalling with the word?

Phillip: Well, I don't believe in ghosts per se. But hey, I don't really believe in the veracity of any stories either—no matter how seemingly factual. Like you said, *we* are the voluntary and involuntary creators of stories and our memories (Anderson, 2004; Edensor, 2005)—as much as we are the creators of our journeys, and our ships. So that's what I mean about mysticism: nothing is quite factual—everything is ephemeral, contingent, relational.

Rhys: Ghosts, too.

Phillip: Exactly, ghosts, too. Ghosts are nothing special, nothing spectacular (Edensor, 2008). Like de Certeau (1984, p. 108) said, "haunted places are the only ones people can live in." There is nothing outside of haunted places. In the stories we share about our places and our ferries—but in general in the stories that all people everywhere else share about their places and their technologies of mobility—there are performative characters. These characters play all kinds of roles. James Carey (1989), for example, talked about technology—in general—as playing the role of a trickster in much of the industrialized world's discourse about it. You know how photocopiers tend to break when you badly need them to copy a document? Or how computers crash right before you've had a chance to back up your files? In the same sense technologies of spatial mobility seem to love messing with our relationship with travel—making it feel routine-like and uncertain at the same time.

Rhys: So they're a particular kind of ghost, in a way, not the uncanny or scary type, but more like...

Phillip: More like a trickster. Like an alchemist, a magician, a wizard. I mean, obviously it depends on the story—I'm sure there are stories in which technologies of mobility—from the Titanic to hijacked airplanes—play dramatically different roles. But BC Ferries, it seems to me, very often play the role of *conjuring* up long forgotten memories (Maddern, 2008). Like many other material objects (Van der Hoorn, 2003) they awaken our memories of "unacknowledged lives" of fellow islanders, of "old habits and dead routines" (Moran, 2004) from their slumber and animate our relationship with our past. Like "intrusions" into the mundane they "reveal the powerful affective, sensual and imaginative effects" of "vestiges" of the past, "which, like the figure of the ghost, are amorphous and slippery" (Edensor, 2008, p. 314). In writing about cities Amin (2004, p. 38) said that they "must be summoned up as temporary placements of ever moving material and immanent geographies, as hauntings of things that have moved on but left their mark." And that for me applies so perfectly to ferry boats, and to all technologies of spatial mobility in general. With all the present emphasis on the mobile, the hauntings of mobility technologies remind us of the material power of obduracy.

Rhys: So, conjuring, for sure. But enchanting, too, right? Think of the bond we build with them, as in the story of people coming back from a long trip.

Phillip: Absolutely. Like art and technology, ghosts have an enchanting function, a dissimulative power which makes things appear for what they are not, confusing the lines between being and not being, past and present, absence and presence (Gordon, 1997; Holloway & Kneale, 2008). Ferries, for example, are not our homes. Yet so many of us feel as though they were. We make love on them, we cuddle up and go to sleep on them, we consume meals and play together on them, we often find the same kind of bodily comfort we feel in the comforts of our own home in them, so it makes sense to also feel at home on them after a long trip—well before walking on the doorstep.

Rhys: Conjuring, enchanting, what else?

Phillip: Hexing, too. There are times when boats themselves seem the victim of a hex, like the *Queen of Victoria* and the *Queen of the North*. But they themselves are capable of hexing in turn. I sure felt that way right after moving to Gabriola, for whatever reason, I kept on missing ferries. I hadn't missed a sailing in four years of fieldwork, and after becoming dependent on the ferries I missed ten in a month.

Rhys: I guess in general you can say that their power—regardless of whether it unfolds through conjuring, enchanting, hexing, or whatever a story may reveal—is transformative, right?

Phillip: Exactly. The key function of performance is transformation (Schechner, 2003). Technologies of mobility transform space, travellers, and everything in between. Their key power is haunting, in a way, because we do not immediately recognize their transformative agency. Most of us view technologies of mobility as deprived of agency—as having nothing but the mere instrumental functionality we endowed them with. So in a sense, our shortsightedness, our inability to understand unintended consequences and our inability to understand the totality of relationships they are embedded in, is the very origin of their "animation." They are ghosts who haunt us because we ignore them for who they are, because we have given them birth without acknowledging their existence.

Rhys: Hence their effervescence?

Phillip: Yeah. For me the effervescence of the haunting power of technologies of mobility is a bit of a genie-in-a-bottle type of phenomenon. It's the energy with which "the debris of shipwrecked histories" "raise up" and "burst forth" to our conscious awareness (de Certeau & Giard, 1998, p. 138).

Our collective ignorance about the way in which technologies are animated, through the multiple relationships we have with them and they have with us, is a bit of a pressure cooker, you know. The more we ignore the transformative power of technology, and the more we turn a blind eye to unintended consequences and multiple roles, and the more we relegate technology's agency to the stigmatized category of anthropomorphism, the more we jam energy into the pressure cooker. All this energy builds up and finds performative release through all the narratives we share with one another. The more we ignore it wilfully, the stronger it comes back to haunt us—and the stronger its various transformative powers.

Rhys: So, what exactly is effervescence for you, then?

Phillip: A measure of the haunting potential of a technology. The more we ignore its power by taking it for granted, the more it bubbles up, and the more vivacious it acts when left unconstrained through the stories we create and share with others.

Rhys: You know, in a way this might explain why all of our ferry stories, the ones we share and know, are always met with a common response when we hear them and tell them to each other. We always respond with a burst of recognition—'Yah, I know that!'

Phillip: True. We tell each other stories we're familiar with, stories that contain characters we've seen or heard of—but never fully acknowledged, like a ghost of sorts.

Rhys: It is pretty well established these days that discourse is *performative* in the senses that I mentioned above (Butler, 1993). Discourses are not just words, but rather the enabling but dissimulative mechanism that creates what is possible and what is not possible, and makes it look natural. But there is more to it than just this, I believe. And the discursive formations they create exist not simply out of an accretion of experience and norms, but also out of a set of organic potentialities present inside us. By this I am thinking of the kind of psychological work of Carl Jung (1959) when he talked about archetypes. For Jung, archetypes were the manifestations of possibilities present within all of us. And some of these ferry stories address this by making potentialities which we recognize within ourselves occur in a story which happens to someone else. Let me give you an example.

Back when I first moved to the island in the early 1970s, the Canadian Pacific ferries used to still run between Vancouver and Nanaimo. The *Princess of Vancouver* was an old boat which carried the traces of an earlier age of coastal steamers. The journey was long, but went from harbor to harbor, with convenient times. It had a bar and a good restaurant. And it was small compared to the BC Ferries running even then. It had a relatively small outside deck, which would fill up during the daylight trips. The crew made their own spaces for relaxation during coffee and lunch breaks, and one of these was the provision of three leather arm chairs, liberated from the bar, which sat up on the roof of the top passenger cabin right behind the funnel. Well, I used to sneak up the metal ladder and sit in the chairs when there were no crew members around. Being a regular, when a member of crew came up I would smile and leave them to it. I have such great memories of sitting on the very top of the boat, almost like I was steering it. I sat alone, unable to see anyone else, with the wake streaming out behind me and the destination out in front. It is a memory I will never forget. It is much like that scene in the movie *Titanic* where the two main characters stand at the bow with the wind in their hair and say something like, "I am the King of the World!" That was how I felt. And when I tell this story, everyone who hears it can relate to it, can put themselves right there. It is why that scene in the film is so memorable, and why my story elicits such a response when I tell it. It is a discursive formation of mobility, of our dream to move, to control our movement, and a performance of the transcendental liminality of that mobility.

So we have come full circle, Phillip. From movement, to home, and back again to movement. The cold metal boats and the warm lives which fill them have a reality that emerges out of their mutual encounters. What we call, *BC Ferries*; what we call, *Islanders*; these places we call, *The Islands*—they all take the specific realities they do only in their intersection, in their *enactment*. But that enactment isn't preordained. It's a matter of mundane encounters, but also the result of a constellation of deep currents which flow both below and across the surfaces of our lives, bubbling up in momentary meanings which are effervescent, yet which somehow adhere to become the realities we recognize and share. We are haunted by the sights, the sounds, the smells, the feelings they engender. And in the movement lies the stability. Which is why ferries are "good to think with" whether you are a social scientist or an old truck driver passing the time with some buddies until the solo part of your journey begins again.

Phillip: I know what you mean, Rhys. Ferries, in a way like all technologies of mobility, are the ghosts of those dreams—characters outside of our control who, through their passages into place and in and out of our lives, "emerge to rekindle the past through unexpected confrontations with sights, sounds, smells, and atmospheres" (Edensor, 2008, p. 327). And, love them or hate them, like dreams and nightmares, we can't get rid of them; we can only learn to live with them as the material media, and the performative mediums, which transport us to lifeworlds we couldn't reach on our own.

Notes

1. BC Ferries is a monopolistic, privately owned, but publicly subsidized and overseen company. It ranks as one of the world's largest ferry transportation companies with annual revenues in excess of CAD$640 million and a fleet of some three dozen vessels of varying size. BC Ferries conducts operations in 47 ports of call through a multitude of routes ranging from 10 minutes to 36 hours in length. Because there are no bridges competing with BC Ferries, these boats are considered to be lifelines. People use the ferries for different reasons. For some the ferries are the door to the outside world, as only a small grocery store and a post office may be available in their communities. For others, ferries are the way to reach supermarkets, supply centres, specialty stores, health, education, administrative centres, workplaces, family members and friends, airports, or leisure destinations. Some people may travel as frequently as almost every day (as it often happens in the Gulf Islands) or as infrequently as once, twice, or half a dozen times a year (as it often happens on Vancouver Island, the North and Central Coast, and the Sunshine Coast).

2. Fieldwork formally began in the spring of 2006 and ended in the summer of 2009. Data collection unfolded through repeated travel to each of the communities served by BC Ferries via regularly scheduled ferry service, for a total of over 250 trips. The fieldwork also comprised about 400 qualitative, semi-structured interviews with a diverse sample of island and coastal residents. More information on the research procedure can be found in Vannini (2011).

Acknowledgment

Phillip Vannini would like to acknowledge the funding assistance of the Social Sciences and Humanities Research Council of Canada, which made possible the research upon which this chapter is partly based.

References

Amin, A. (2004). Regions unbound: Towards a new politics of place. *Geografiska Annaler, Series B: Human Geography, 86*(1), 33-44.

Anderson, B. (2004). Recorded music and practices of remembering. *Social & Cultural Geography, 5*(1), 3-20.

Butler, J. (1993). *Bodies that matter: On the discursive limits of 'sex.'* New York, NY: Routledge.

Carey, J. (1989). *Communication as culture: Essays on media and society.* Boston, MA: Unwin Hyman.

de Certeau, M. (1984). *The practice of everyday life.* Berkeley, CA: University of California Press.

de Certeau, M., & Giard, L. (1998). Ghosts in the city. In M. de Certeau, L. Giard, & P. Mayol (Eds), *The practice of everyday life, Volume 2: Living & cooking* (T. J. Tomasik, Trans., pp. 133-143). Minneapolis, MN: University of Minnesota Press.

Edensor, T. (2005). The ghosts of industrial ruins: Ordering and disordering memory in excessive space. *Environment & Planning D, 23*(6), 829-849.

Edensor, T. (2008). Mundane hauntings: Commuting through the phantasmagoric working-class spaces of Manchester, England. *Cultural Geographies, 15*(3), 313-333.

Gordon, A. (1997). *Ghostly matters: Haunting and the sociological imagination.* Minneapolis, MN: University of Minnesota Press.

Holloway, J., & Kneale, J. (2008). Locating haunting: A ghost-hunter's guide. *Cultural Geographies, 15*(3), 297-312.

Jung, C. G. (1959). *The archetypes and the collective unconscious* (Vol. 9). New York, NY: Bollingen Foundation.

Maddern, J. F. (2008). Spectres of migration and the ghosts of Ellis Island. *Cultural Geographies, 15*(3), 359-381.

Maddern, J. F., & Adey, P. (2008). Editorial: Spectro-geographies. *Cultural Geographies, 15*(3), 291-295.

Moran, J. (2004). History, memory and the everyday. *Rethinking History, 8*(1), 51-68.

Schechner, R. (2003). *Performance theory.* New York, NY: Routledge.

Thrift, N. (1999). Steps to an ecology of place. In D. Massey, J. Allen, & P. Sarre (Eds.), *Human geography today* (pp. 295-322). Cambridge, UK: Polity Press.

Van der Hoorn, M. (2003). Exorcizing remains: Architectural fragments as intermediaries between history and individual experience. *Journal of Material Culture, 8*(2), 189-213.

Vannini, P. (2008). A queen's drowning: Material culture, drama and the performance of a technological accident. *Symbolic Interaction, 31*(2), 155-182.

Vannini, P. (2011). *Ferry tales: Mobility, place, and time on Canada's west coast.* New York, NY: Routledge.

Part III:
Technology, Technicians, and Mobility in Everyday Life

11

Mobile Phones as a "Necessary Evil": Canadian Youth Talk About Negotiating the Politics of Mobility

Tamara Shepherd and Leslie Regan Shade

In research conducted with young people on their use and perceptions of the mobile phone, the Pew Internet and American Life Project argued that understanding how youth use mobile phones is vital to creating effective policy based on the reality of how the technology is used (Lenhart, Ling, Campbell, & Purcell, 2010, p. 10). The ways that young people use their mobiles is in turn often influenced by the policies and pricing schemes of wireless telecommunications companies, which forms the backdrop for this chapter. In order to examine how mobile phones are marketed toward and perceived by young Canadians, we adapt a framework developed by Mackay and Gillespie (1992), who argued that inserting a cultural studies approach into a social shaping of technology (SST) perspective allows for a nuanced examination of three interrelated spheres of technology: regulation, marketing, and appropriation by users. These spheres further relate to Cresswell's (2010) "constellations of mobility" framework, which is attentive to historically contingent iterations of mobility as "movement, meaning and practice and the ways in which these are interrelated" (p. 26). In this chapter, we relate how a small group of young Canadians discuss these constellations of mobility in their everyday lives, particularly in terms of how their appropriation of mobile devices is shaped by regulation and marketing discourses that position them as mobile consumers.

The first sphere, regulation, is examined through an overview of the current contested state of the wireless industry in Canada, wherein debates over levels of foreign ownership in the telecommunications sector revolve around competition, affordability, and cultural sovereignty. Several mobile marketing campaigns by incumbent firms and new upstart companies specifically targeting youth are analyzed to comprise the second sphere. The attitudes and practices surrounding the everyday use and economics of the mobile phone by a selected group of Canadians aged 20–24 comprise the third sphere. When the discussions of young people's everyday interactions with mobile devices and culture are placed in dialogue with their positioning within regulatory and marketing discourses, the constellations of mobility that comprise these spheres are expressly political. As Cresswell contended, constellations of mobility entail specific politics of mobility, based on the regulation and control of movement within broader architectural, social, and governmental structures.

This research is timely given the opening of the wireless spectrum market in Canada, where new entrants are competing with the incumbent carriers amidst marketing campaigns that are particularly targeting young Canadians. While mobile phone adoption has become common—99% of the population subscribes to mobile services—the younger demographic is seen as a key site for emerging trends in wireless communication (CWTA, 2010). By exploring the practices of young people within the broader context of the national wireless industry and its marketing strategies, this chapter aims to highlight how the economics of youths' mobile phone use might impinge on the broader politics of mobility in relation to young people as both consumers and political actors in the context of Canada's current telecommunications regime. As the historical context where certain constellations of mobility take shape, we frame Canadian telecommunications politics as a space where the meanings of mobility are being defined. In examining how young mobile phone users are represented through discourses of regulation, marketing, and user appropriation, we extend the framework of constellations of mobility as a way of interconnecting young people's everyday experiences of this increasingly prevalent technological sphere.

Regulation: Shaping the Mobile Terrain

Canada's wireless industry, dominated as it is by three established telecommunications firms—Rogers Communications, Inc.; BCE's Bell Mobility, Inc.; and Telus Communications Company—reflects both the constraints of mobile phone technology and the legislative parameters of the Federal government's telecommunications policies. According to a 2010 report, these "big three" players account for 95% of the Canadian market and enjoy the highest profit margins of any wireless corporations in the developed world (Nowak, 2010). Part of this dominance has to do with the physical exigencies of Canada: providing wireless infrastructure to a relatively sparsely distributed population across a vast land area is challenging, and only larger conglomerates are able to exercise country-wide network coverage (Senate Canada, 2010).

Regulatory regime

Along with the infrastructural parameters that have contributed to the dominance of the big three corporations, regulatory restrictions on foreign ownership have prevented major, global competitors from entering the market. The Canadian Radio-television and Telecommunications Commission (CRTC), the country's central regulatory body, has established and policed rules on foreign ownership and control of broadcasting and telecommunications services since its formation in 1976. Traditionally, these ownership restrictions applied to companies that were not as profitable as the big three wireless providers. The big three's 2007 revenues of approximately USD$440 per capita are high when compared against the revenues of other wireless industries in developed countries (e.g., USD$400 per capita in Korea; USD$390 per capita in the US; and USD$280 per capita in Sweden (OECD, 2009, figures 3.5 and 3.8). Public dissatisfaction with the way that Rogers, Bell, and Telus have exploited their market share and charge the third highest prices of all 30 OECD member countries, has led the CRTC to consider strategies for increasing competition between wireless service providers (OECD, 2009; Senate Canada, 2010).

In 2008, Industry Canada held an Advanced Wireless Services spectrum auction that reserved the 40 MHz portion of the electromagnetic spectrum—one of a series of wavelengths that carry wireless signals—for licensing to new entrants to the market (Senate Canada, 2010). The auction generated over $4bn in profit for the federal government and ended with the big three companies retaining and expanding their spectrum licenses, while new entrants (Shaw, Quebecor-Videotron, Bragg, DAVE Wireless, Public Mobile, and Globalive Wireless) were able to bid for a share of the 40 MHz spectrum (Canadian Press, 2008).

This resulting shift in the landscape of wireless providers in Canada reflects the federal government's emphasis on wireless communications and the promotion of a "digital economy" (Sawchuk & Crow, 2010). By holding spectrum auctions, the government set up its recommendations for a "national digital strategy" to involve the liberalization of foreign ownership restrictions in telecommunications (Government of Canada, 2010). As Industry Minister, Tony Clement, proposed in his 2010 announcement of consultations on foreign ownership, liberalization will take shape in one of three forms: removing all restrictions; increasing the limit of foreign investment from the current 20 to 49%; or lifting restrictions for carriers with less than 10% of market share (CBC News, 2011). By allowing increased foreign ownership, the government hopes to attract investment in Canada's wireless infrastructure, claiming that consumers will see benefits in service quality and affordability (Senate Canada, 2010).

Wireless Market in Canada, Winter 2011		
Operator	Service Brand	Subscriber Market Share 2009
Bell Mobility	Northwestel, NMI Mobility, Virgin Mobile, Solo Mobile	30%
Rogers Wireless	Fido, Chat'r	37%
TELUS Mobility	Koodo Mobile, Mike	28%
Public Mobile		Unknown
Globalive	Wind Mobile	Unknown
Mobilicity (formerly DAVE Wireless)		Unknown
SaskTel Mobility		Unknown
Videotron		Unknown

Figure 1: Wireless Market in Canada, Winter 2011 (CRTC, 2010)

These regulatory transitions meant to increase competition in Canada's relatively closed wireless industry show how constellations of mobility—shaped in part by infrastructural parameters determined by government regulation of industry—are contingent both geographically and historically (Cresswell, 2010). Increasing wireless industry competition in Canada has raised tensions between the current Conservative administration that seeks to lift long-standing CRTC foreign ownership restrictions. The story of Globalive's WIND mobile brand offers a salient example of the friction in Canadian regulation of wireless services, with its foreign investment from Egyptian firm, Orascom Telecom Holding SAE, causing a back-and-forth negotiation between the federal government and the CRTC of its status as a "Canadian company" (CRTC, 2009). With the election of a majority Conservative government in May 2011, it is anticipated that opening up foreign ownership restrictions will be a priority for the forthcoming legislative agenda. The outcome of this regulatory struggle will in turn affect the constellations of mobility afforded to individual Canadians who use mobile devices.

What is particularly interesting for us about stories like Globalive's in the context of the wireless industry in Canada is that young people form a crucial market where innovative ownership, control, and pricing structures battle for consumers. For example, Rogers's takeover of Microcell's lower-priced Fido in 2004 served to reinstate the company's dominant market share by bringing rising public anger directed toward the dubious pricing practices of the big three providers (Sawchuk & Crow, 2010). As the Marketplace story concluded, cellphone service actually costs carriers very little, compared to the "astronomical and sometimes mystifying charges" faced by consumers.

In this context of widespread public anger and mistrust, newer entrants to Canada's mobile phone industry have used the high costs and hidden fees of the big three and their subsidiaries as selling points. In its inaugural "Hot Dog Fees" advertisement, for example, WIND mobile focused on this very issue, using humour to highlight the absurdity of mobile providers' hidden, added fees through the analogy of a hot dog vendor demanding extra fees for preparation, buns, napkins, and condiments. Newer providers like WIND and Public Mobile have explicitly made their contracts more transparent and therefore trustworthy. This move, intended to appeal to Canadians frustrated with the high-priced, binding contracts of the big three, also represents a response to the practice of discount brands like Virgin Mobile and Koodo to target younger consumers with "tabs" rather than contracts. The tab system allows consumers to acquire a handset for little or no initial cost, while adding a small portion of that cost to the monthly bill without a fixed-term contract. So while tabs enable consumers to avoid commitment to a high-priced contract, their bills still feature hidden fees as part of the tab system. Other companies seek to attract consumers with credit incentives, such as the "Fido Dollars." Yet tabs and credits can be seen as further complicating the pricing structure of these service providers, who perform a kind of benevolence to consumers in order to obscure unfair fees.

Thus despite the diverse attempts of mobile carriers to trade on trust in their promotional materials, an atmosphere of cynicism seems to pervade the public attitude toward Canadian wireless companies. Globally high prices for wireless services not regulated by the CRTC appear to endure given the uncertain situation of market newcomers. As the only recourse for consumer protection in the mobile phone sector, the independent, not-for-profit Commissioner for Complaints for Telecommunications Services (CCTS) claimed that the price charged by a service provider "is a business decision that it alone is entitled to make. Marketplace competition means that you may find a better deal with another provider" (CCTS, 2011, n.p.). Yet while marketplace competition has so far not resulted in fair and/or lower prices for Canadians, consumers—including youth—have become wary of the promises of mobile phone marketing.

Appropriation: Young Canadians Talk About Mobile Phones

The lack of transparency in Canadian mobile phone advertising, as an extension of the consolidation of the industry in this country, has engendered a deep-seated mistrust of mobile service providers by many young Canadians. For this pilot study, we conducted informal focus groups with students aged 20-24 in Concordia University's undergraduate program in Communication Studies.[1] These in-depth discussions provided a detailed account of the constellations of mobility which are apparent in student attitudes toward mobile phones. Based on a series of open-ended questions, participants were encouraged to discuss their relationships to mobiles in terms of the features of their handsets, their use of the phones, social conventions, advertising, pricing plans, and dealings with service providers. Their reflections acknowledged the embeddedness of mobile phones within an everyday communicative ecology, but were accompanied by skepticism about the advertising, pricing schemes, and customer service of mobile providers. Quotations from selected focus group participants are used in the discussion below to illustrate the way that young Canadians negotiate the mobile phone landscape in Canada.

The indispensable mobile phone: Uses and conventions

Our participants use their mobile phones multiple times throughout the day and, despite the sometimes aggravating expectation to be "always on," mobile devices are seen as being virtually indispensable. This attitude is consistent with recent figures from the Canadian Wireless Telecommunications Association (CWTA) and the Harris-Decima *2008 Wireless Attitudes Study* which respectively reported that over 99% of the country's population is served by wireless coverage and that cell phone penetration is highest among the 18-34 age group (CWTA, 2010; Harris-Decima, 2008). In keeping with these statistics, our participants opined that it was highly unusual to know someone without a mobile phone. As Michelle claimed, such young people are viewed suspiciously: "you feel the anxiety socializing with people that don't have it, because it's just a totally different set of habits and standards to what you're used to." The habits and standards that go along with mobile phone usage for young people came up repeatedly in our discussions with participants, indicating how mobiles have become embedded into everyday social practices.

The embeddedness of mobiles into youth's everyday lives has been studied across a variety of youth cultures, which seem to share some similar features. For instance, Ishii (2006) described how Japanese youth use mobiles more than any other communications technology, facilitated largely by the unique provision of "advanced services" by that country's wireless service providers (p. 349). In Sweden, as Thulin and Vilhelmson (2009) discussed, teenagers similarly show a dependence on relatively affordable mobile communication, preferring to use text messaging over computer-based instant messaging as part of enacting "always on" connectedness. These two studies represent only a portion of the extensive literature on how youth in Nordic and East Asian countries in particular, where wireless services are relatively affordable for young people, use mobile phones to both reinforce existing social practices and enact new kinds of communication (see Ito, 2005; Johnsen, 2003; Katz & Aakhus, 2002; Kim, 2005). The various ways that young people use mobiles worldwide are thus somewhat contingent on the market dynamics within particular countries, even though on average, youth all over the world has been seen as a key constituency of mobile consumers as part of their coming of age and negotiation of identity within mediated social contexts.

Beyond being useful for coordinating meetings, making calls, or feeling safe (the typical practices that are adopted early on by American teens, e.g., Blair & Fletcher, 2011), mobiles perform a variety of functions that the young people in our study said they would have difficulty living without. These functions range from the simple act of changing ringtones or alarm music (as Lubomir did with his Sony Ericsson handset), to maintaining long-distance social ties through BlackBerry's BBM messaging system (as Lara does), to using a translation application on the iPhone to understand concepts presented in class (an example offered by Véronique). Particularly for those with smartphones, the participants were able to list a variety of features and applications that they use on a regular basis, and that they would miss if something happened to their mobiles. As Véronique recalled, finding her way to a recent meeting without the GPS on her iPhone was stressful and unnerving.

The anxiety of being without the phone was something that participants agreed upon; but at the same time, they also expressed more critical attitudes to the phone as a kind of "leash." Michelle described mobiles as a "social norm of protection" for people who feel the need to be occupied at all times and protected from interacting with strangers: "People are so awkward with themselves if they are just left to their own devices in public." Moreover, as all the participants noted, the expectation to be always available was another way that mobile phones served as leashes. To combat this expectation, Lubomir would deliberately leave his phone at home when going out cycling, and Lara would not take her phone on hikes. Michelle would leave her phone on the vibrate function throughout the day, and then return missed calls at her convenience. As Véronique asserted, "you're not obliged to always be available." Lara too agreed with these sentiments, claiming to want to rebel against the always-on expectation, but she also professed her "addiction" to the phone: "I really envy people who can not be on their phone..., but I need to take it when I go to school, because I need to do something on that shuttle bus." Even during the focus group meeting, Lara communicated with friends over BBM, while telling the story of feeling a "phantom phone vibration" after having lost her previous iPhone. The juxtaposition of these two themes—of the phone's indispensability and the simultaneous pressure of being always available—revealed the group's conflicting feelings about mobiles.

Perhaps most indicative of the way that the participants work to negotiate the habits and standards of mobile phone use, a large portion of the focus group time was spent on discussing social norms around texting. According to the Harris-Decima study, the frequency of text messaging nearly doubled between 2006 and 2008, with the highest percentage of texters aged between 13 and 34 (Harris-Decima, 2008). Yet in that study, talking was still reported to be the "most important" function of the cell phone for all age groups (p. 15). Reflecting the trends found in more recent research, our focus groups revealed that "cell-phone texting has become the preferred channel of basic communication" for youth (Lenhart et al., 2010, p. 2). Texting was the most common means of mobile communication among young people and was, paradoxically, viewed as being both less and more intrusive than calling. So while texting was seen as the easiest way to set up meetings or have more private conversations in public spaces, it was also framed as disruptive and rude. As Michelle advised, even though texting has become such an entrenched practice, "when someone's talking, at least say 'excuse me.'" Discussions like this one around text etiquette led many of the participants to claim that calling was more to-the-point and therefore less intrusive, but at the same time, they acknowledged that texting might be more appropriate than talking in places like public transit or at the hair salon. The social norms around texting proved to be an important theme for participants to express in terms of the double standards (as Lara confirmed, it's annoying when other people are texting on their phones all the time, but it's difficult to not do that yourself) and contradictory attitudes towards cell phone use.

Service providers: From bad to worse?

Another aspect of the conflicting attitudes that came up during our discussions of texting involved the pricing plans of wireless carriers. The fact that texting was often more affordable than calling also contributed to its widespread use among the participants and their friends. Given the lack of substantial competition and globally high prices among the country's wireless industry players, our group of young Canadians had reason to be concerned with mobile phone pricing. A 2010 Harvard University report on international rates of broadband access concluded, for example, that Canada was a weak performer in terms of Internet speed, prices, and 3G mobile penetration (Benkler et al., 2010). Unlike the prevalence of prepaid calling plans offered in many European and Asian countries, or the "low nationwide flat-rate pricing structure for mobile telephony" in the US—both of which allow youth to make calls affordably (Ling, 2007, p. 62)—talk time in Canada tends to be expensive through the overly complicated and often deceptive pricing structures of Canadian wireless providers. As the shape of the industry changes, however, this gap may narrow with new entrants offering more affordable prepaid phone plans.

In the meantime, the overall feelings that our participants expressed about wireless providers were mostly negative, as they shared stories of poor customer service and convoluted contractual terms. In fact, as Lubomir contended, bad news stories about wireless carriers come up on a regular basis among his peers, where "people only talk about it if they're complaining; there's nothing good to say about it."

In terms of customer service, the group agreed that calling companies to complain, to the point of threatening to switch carriers, was the only way to receive better service. For some of the participants, these dealings with service providers were still taken care of by their parents; for instance, Lara noted that her mom deals with billing and service complaints to Telus about their family plan. Lara described how her mom handled the rest of the family's service issues by leveraging the whole family's participation in the plan. Similarly, Michelle discussed her former family plan subscription to Bell Mobility, from which she withdrew because the billing was so convoluted and because in general, "everyone hates Bell." Now Michelle has a mobile phone from Virgin, and despite initially seeming "more accessible and easy," her experiences with customer service have been frustrating: "every representative had something different to say." Regardless of the carrier, it appeared that the participants almost expected that customer service would be a hassle. As Lubomir said, although neither he nor his family have had major issues with Fido, his girlfriend had been swindled into paying for repairs to her Telus handset that she did not need to operate the phone: "basically they screwed her."

The sense that mobile providers are quick to "screw" customers was shared by the group, who felt that compared to carriers in other countries, Canadian wireless companies were untrustworthy. Lara described how Australian cell phone plans were much less "convoluted," and Véronique noted that she could easily top up her pay-as-you-go phone in Paris for five euros at a time. Michelle recounted her boyfriend's experience of converting from a German phone, which the Fido representative claimed was as simple as purchasing a plan and new SIM card on their network. Yet after he had paid for the SIM card and signed the new contract, the company informed him that there was a "problem with the connection," meaning he would need to purchase a new Fido handset all together: "It's like, is that manipulation? Like, 'oh you have to buy our phone.' Or is that actual?" Michelle's doubt about the veracity of the company's claim led her and Véronique to frame the dubious practices of wireless providers in terms of their status as "platforms"; consumers prefer to spend their money in one place through bundled services with single providers, but these providers are conglomerates that own several subsidiaries—subsidiaries with different names, obscuring the structure of corporate ownership for consumers. As Lubomir and Lara concurred, the wireless industry is "basically a monopoly" in Canada.

Lack of transparency in mobile advertising

Armed with this sense that Canada's wireless providers are essentially untrustworthy, the participants approached the advertising campaigns with a fair amount of skepticism. Particularly concerning the ads' promotion of pricing plans, the group confidently asserted that they knew the figures quoted were always a gross underestimation. Lubomir even offered the formula that one should expect the cost to be 50% higher than the figure cited in the commercials. The participants seemed to share the attitude that the cost of phones was inevitably greater than the price quoted in the ads, though they did not claim to pay much attention to the ads themselves. As Véronique said, "When I buy a new phone, I'm not thinking, 'oh these ads look nice, what do they have to offer?' I'm just going to figure out how much it costs. For the iPhone, I saw a student plan for example." Yet despite claiming that the ads were not influential on their choices of mobile service providers, the group did agree that pricing plans featured in the ads were misleading and overly complicated.

Michelle's experience with Virgin Mobile was exemplary of the way that the advertising and pricing plans target younger consumers in particular, while ultimately misleading them into expensive services. Its seemingly uncomplicated pricing plans drew Michelle to the company, although she was unimpressed with its racy ad campaign, including a billboard of a young woman lying on a couch with the double entendre tagline "c'est meilleur avec Virgin" ("it's better with Virgin") that looms over Concordia's downtown campus. Regardless of the sexiness and, as Michelle put it, "MTV generation" ethos of Virgin's ads, they offered a $30 per month unlimited texting plan that fit her need for a no-frills handset with basic functionality. Yet even though she signed up for this lower-cost plan with a basic handset, Michelle's bill tended to be consistently higher than advertised: "I feel like whatever plan I get, I'm always going over, and it's always like 70 or 80 dollars, and I just get so pissed with it. It's so stupid, it's such a little piece of crap." Her frustration at Virgin's misleading pricing scheme was matched by an equally exasperated attitude about the company's customer service, which simultaneously panders to youth through its casual tone while never satisfactorily explaining the extra charges accrued with each bill. Her experience with Virgin has led Michelle to be eager to switch providers once her contract expires in another year.

The other participants reported similar feelings of frustration in how advertised pricing plans seemed deliberately misleading, or at least, as Michelle surmised, "constantly changing; I feel like we never really know what's actually happening." Véronique's Rogers plan for the iPhone, for instance, was offered through the Concordia University website as a special student deal. While the context of the University's website lent credibility to the offer, its initial three months of unlimited usage ended up feeling manipulative to her: "you get used to being everyday on the Internet, and then you get your habits, and then, Surprise! Month number four, like, oh my god. I have to stop doing that." Moreover, such abrupt changes to service costs are almost never clarified by a call to customer service; in Véronique's case, she was told by Rogers representatives that her bill was not "in the system," and so could not be explained; and, once it was in the system, they justified the charges by saying, "oh don't worry, everybody has that." Lara also reported feeling manipulated by hidden charges through Telus's misleading description of its My5 feature. She had successfully entered her five friends' phone numbers online, and only after incurring significant charges while using the supposedly free program, was informed that the My5 feature didn't cover numbers outside of Canada: "That screwed me over. It's like, 'just do it online,' and so I just did it. They cheated me."

Lara's sense of feeling cheated provided an apt description of the participants' attitudes toward mobile phone advertising. While the ads claim to offer affordable pricing plans and special deals for younger customers, the young people in this group were well aware of the fact that, as Lubomir said, "the ads are completely not in synch with what it is on paper." Especially after the participants had seen what mobile phone service was like in other places, including the US, Australia, and European countries, they felt that despite their promotional strategies for low pricing, Canadian wireless companies were price gouging consumers. Because of this perceived unfair pricing, Lubomir mentioned that he hasn't purchased $10-per-month call display as part of his plan: "It's not worth $10. It is important, but I mean that's a little ridiculous. Just because they know that everyone wants it, I feel like they're jacking the price up." Lara agreed, saying that Canadian wireless carriers charge unfair prices "because they can," obscuring these charges in assorted hidden fees. The incongruity between the plans advertised in mobile providers' advertisements and the cost of the actual monthly bill resulted in participants' feeling exasperated with the way that wireless service in Canada is, as Lara characterized it, "just not straightforward at all."

Consensus: Mistrust

The participants' general mistrust of mobile phone providers, articulated throughout the focus groups, led them to develop various ways of negotiating the Canadian wireless landscape. When initially choosing a provider, for example, participants claimed that because the advertising was perceived as misleading, they tended to seek word-of-mouth advice. "People do that first," Michelle noted, "because they don't trust them [wireless companies]. I don't think anyone does. Everyone's really cynical about it I think, but you have to do it." To this, Véronique chimed in, "It's true, no one trusts them!" So Michelle went with a friend's advice to try Virgin Mobile, and Véronique signed up with Rogers through the Concordia University student plan. These decisions were seen as temporary, however, as the young people agreed that switching providers was a commonplace practice; they reported anticipating the end of their current contracts to be able to try out another company.

As such, even once they had decided to deal with a particular provider, the feeling of mistrust remained. As Lara and Lubomir had discussed, price gouging seemed to be an inevitable practice of Canadian wireless carriers. And so the participants had each cultivated usage strategies to cope with the high costs of mobile service. Balking at the $10-per-month charge for call display with Fido, Lubomir declined to purchase the service, choosing instead to simply pick up the phone to find out who is calling: "If someone is calling, I *must* pick up." Similarly, since she has a BlackBerry, Lara detailed how she prefers to use the free BBM messaging system over regular texting, noting how she communicates more often with her contacts who are also on the BlackBerry system since it incurs no extra charge. Michelle's strategy for keeping her costs low, although not always effective, was to have an older handset with fewer features. These everyday ways of negotiating high costs for cell phone service offer an example of how constellations of mobility—comprised of movement, representation, and practice—get shaped by the institutional regulation of mobility (Cresswell, 2010, p. 27). This group of young people's cynical attitudes toward Canadian wireless service providers' advertising and pricing schemes contribute to their negotiated appropriation of mobile communication.

Conclusion: Young Canadians and the Politics of Mobility

As discussed by the participants, mobile device appropriation happens within interconnected constellations of mobility that incorporate complex meanings and politics of contingent "formations of movements, narratives about mobility and mobile practices" (Cresswell, 2010, p. 17). In applying the constellations' paradigm to young people's relationships with mobile technologies, the meanings of mobility take shape within the interrelated spheres of regulation, marketing, and appropriation by users (Mackay & Gillespie, 1992). Given the current landscape of Canadian wireless service providers, the big three incumbent carriers and their subsidiary brands were viewed by the young people in our study with a significant degree of cynicism. As a kind of "necessary evil," our participants claimed that wireless carriers engage in deliberate price gouging, with their promotional materials obscuring hidden fees to the point where, as Véronique said, "I always feel that there is something under that is manipulating." The newer entrants to Canada's wireless marketplace since the main spectrum auction in 2008 have played on these public feelings of mistrust in their advertising campaigns. Yet regulatory hurdles, along with a lack of word-of-mouth support (as noted by our participants), have made the new entrants' task to break into the Canadian wireless market not without its challenges—constituting a tenuous terrain in which to negotiate the politics of mobility.

For young Canadians reliant on wireless service, the lack of competition in the industry has served to circumscribe their means of appropriating mobile technology, alongside inviting a cynical attitude toward the depictions of youth in the misleading advertising campaigns of service providers. This climate forms an uncertain backdrop for the way that younger users of mobile phones constitute an increasingly crucial demographic in terms of marketing, since they represent a particularly active group of mobile users. The young Canadians we spoke with conveyed an enthusiasm about mobile communication, often seeing their devices as extensions of themselves despite the drawbacks of perpetual availability. In seeing their mobile phones as inevitable technologies of everyday life, they confirmed the broader trend for young people to eschew landline communication (Lenhart et al., 2010, p. 45). In their move away from the established landline infrastructure in Canada, this group of young people has transposed the expectation of an affordable communications infrastructure onto wireless service, and this is the locus for their frustrations about the opaque and unfair pricing plans offered to mobile users.

Acknowledgements

This research was conducted with funding from the Social Sciences and Humanities Research Council of Canada, as part of the "Young Canadians, Participatory Digital Culture and Policy Literacy" project.

Notes

1. Students were recruited through a group email sent to former members of a Fall 2010 course in the Communication Studies department. Sessions were audio recorded and transcribed into word processing software. Participants consented to their first names being cited.

References

Benkler, Y., Faris, R., Gasser, U., Miyakawa, L., Schultze, S., Baudry, J., & Wilks, A. (2010). *Next generation connectivity: A review of broadband Internet transitions and policy from around the world*. Cambridge, MA: Berkman Center for Internet and Society, Harvard University. Retrieved from http://cyber.law.harvard.edu/pubrelease/broadband/

Blair, B. L., & Fletcher, A. C. (2011). 'The only 13-year-old on Planet Earth without a cell phone': Meanings of cell phones in early adolescents' everyday lives. *Journal of Adolescent Research, 26*(2), 155-177.

The Canadian Press. (2008, June 24). Bidding in Canada's wireless spectrum auction hits $4B. Retrieved from http://www.cbc.ca/news/technology/story/2008/06/24/spectrum-four-billion.html

CBC News. (2011, February 15). Globalive ruling to be appealed by Ottawa. Retrieved from http://www.cbc.ca/news/business/story/2011/02/15/clement-globalive-appeal.html

CCTS (Commissioner for Complaints for Telecommunications Services). (2011). FAQs. Retrieved from http://www.ccts-cprst.ca/faqs

Cresswell, T. (2010). Towards a politics of mobility. *Environment and Planning D: Society and Space, 28*(1), 17-31.

CRTC (Canadian Radio-television and Telecommunications Commission). (2009, October 29). *Telecom decision CRTC 2009-678*. Retrieved from http://www.crtc.gc.ca/eng/archive/2009/2009-678.htm

CRTC (Canadian Radio-television and Telecommunications Commission). (2010). *CRTC Communications monitoring report 2010*. Retrieved from http://www.crtc.gc.ca/eng/ publications/reports/policymonitoring/2010/cmr.htm

CWTA (Canadian Wireless Telecommunications Association). (2010). *Subscriber stats for September*. Retrieved from http://www.cwta.ca/CWTASite/english/facts_figures_downloads/SubscribersStats_en_2010_Q3.pdf

Government of Canada. (2010, March 3). *Speech from the throne: A stronger Canada. A stronger economy. Now and for the future*. Retrieved from http://www.speech.gc.ca/eng/media.asp?id=1390

Harris-Decima. (2008, September 12). *2008 wireless attitudes study*. Retrieved from http://www.cwta.ca/CWTASite/english/pdf/DecimaStudy_2008.pdf

Ishii, K. (2006). Implications of mobility: The uses of personal communication media in everyday life. *Journal of Communication, 56*(2), 346-365.

Ito, M. (2005). Mobile phones, Japanese youth, and the re-placement of social contact. In R. Ling, & P. Pedersen (Eds.), *Mobile communications: Re-negotiation of the social sphere* (pp. 131-148). London, UK: Springer.

Johnsen, T. E. (2003). The social context of the mobile phone use of Norwegian teens. In J. E. Katz (Ed.), *Machines that become us: The social context of personal communication technology* (pp. 161-170). New Brunswick, NJ: Transaction.

Katz, J. E., & Aakhus, M. (Eds.). (2002). *Perpetual contact: Mobile communication, private talk, public performance*. Cambridge, UK: Cambridge University Press.

Kim, J. (2005). An examination and comparison of mobile phone uses by adolescents and adults. *Korean Journal of Journalism & Communication Studies, 49*(3), 262-286.

Lenhart, A., Ling, R., Campbell, S., & Purcell, K. (2010, April 20). *Teens and mobile phones*. Pew Internet and American Life Project. Retrieved from http://www.pewinternet.org/ Reports/ 2010/Teens-and-Mobile-Phones.aspx

Ling, R. (2007). Children, youth, and mobile communication. *Journal of Children and Media, 1*(1), 60-67.

Mackay, H., & Gillespie, G. (1992). Extending the social shaping of technology approach: Ideology and appropriation. *Social Studies of Science, 22*(4), 685-716.

Nowak, P. (2010, July 19). Canadian wireless firms still tops in profit: Report. *CBC News.* Retrieved from http://www.cbc.ca/news/technology/story/2010/07/19/canada-wireless-profit.html

OECD (Organisation for Economic Co-operation and Development). (2009). *OECD communications outlook 2009.* Retrieved from http://www.oecd.org/document/39/0,3746,en_2649_34225_48223143_1_1_1_1,00.html

Ovum Consulting Group. (2010). *The benefit of the wireless telecommunications industry to the Canadian economy.* Retrieved from www.cwta.ca/CWTASite/english/pdf/OVUM_Study.pdf

Ross, R. (2004, September 21). Rogers fetches Fido for $1.4 billion. *Toronto Star*, p. D01.

Sawchuk, K., & Crow, B. (2010). Talking 'costs': Seniors, cell phones and the personal and political economy of telecommunications in Canada. *Telecommunications Journal of Australia, 60*(4), 55.1-55.11.

Senate Canada. (2010, June). *Plan for a digital Canada.* Retrieved from http://planforadigitalcanada.ca/images/stories/pdf/report.pdf

Thulin, E., & Vilhelmson, B. (2009). Mobile phones: Transforming the everyday social communication practice of urban youth. In R. Ling, & S. Campbell (Eds.), *The reconstruction of space and time: Mobile communication practices* (pp. 137-158). New Brunswick, NJ: Transaction.

12

A Sociology of Traffic: Driving, Cycling, Walking

Jim Conley

In 1968, Norbert Schmidt-Relenberg (1968/1986, p. 121) wrote that "an analysis of traffic can enrich sociological theory." Little became of his hope, but, in the interim, the "mobility turn" in the social sciences has transformed our understanding of all forms of movement, including automobility. This chapter develops a sociology of pedestrian, bicycle, and automobile traffic by considering driving, cycling, and walking as socially interactive mobilities. The complex coordination of movements between the multiple independent units that make up car, bike, and foot traffic are performed by and between the mobile units themselves, largely through looking at each other. In contrast to the "eye in the sky" view of traffic planners and engineers, for whom traffic movements are akin to flows of particles (Lynch, 1993), and to studies of the physical, emotional, and aesthetic experiences of individual drivers, cyclists, and pedestrians, the relational approach developed here shows how the material characteristics and speed of these pedestrian, bicycle, and automotive assemblages provide opportunities for and limitations on visual contact and social interaction between them.

Starting from Simmel's reflections on the "sociological eye," I develop a typology of mobile looking. Turning then to Goffman, I extend his understanding of traffic as a social order by examining in detail how walking, driving, and cycling as forms of mobile interaction are affected by the material characteristics and speed of the mobile units.

Simmel: The Sociological Eye

In his influential essays, "The Stranger," and "The Metropolis and Mental Life," Georg Simmel (1903/1971a, 1908/1971b) examined how urbanization and modern mobilities affected social life in the nineteenth century. In this context, his remarks on the sociological function of the eye have often been cited in the mobilities literature, notably by John Urry (2004, p. 30; 2006, p. 21; 2007, p. 24). Their significance has not been fully explored, however. Simmel (1908/1969) stated:

> Of the special sense-organs, the eye has a uniquely sociological function. The union

and interaction of individuals is based on mutual glances....This mutual glance between persons, in distinction from the simple sight or observation of the other, signifies a wholly new and unique union between them. The limits of this relation are to be determined by the significant fact that the glance by which one seeks to perceive the other is itself expressive. By the glance which reveals the other, one discloses himself....What occurs in this direct mutual glance represents the most perfect reciprocity in the entire field of human relationships. (p. 148)

As a way of "knowing" others, sight predominated over hearing in the city, Simmel argued. He explained this in part by "the development of public means of transportation" which put people in "a situation where for periods of minutes or hours they could or must look at each other without talking to one another." This, he claimed, put "social attitudes and feelings upon an entirely changed basis," creating problems of "the lack of orientation to the collective life, the sense of utter lonesomeness, and the feeling that the individual is surrounded on all sides by closed doors" (Simmel 1908/1969, pp. 150–151).

Simmel's examples of modern means of urban mobility were mass, public forms such as buses, trains, and street cars, not independent, self-directed forms such as automobiles and bicycles, and he took walking for granted. But the interactional order produced amongst seated passengers in public transport, assiduously avoiding eye contact (Levine, Vinson, & Wood, 1973), is not the same as the mobile independent coordination of action that goes on in car, bicycle, or pedestrian traffic. Walking, driving, and cycling also differ from each other, and the task here is to describe those differences and their implications. Although Simmel considered mutual glances within moving vehicles such as train cars rather than between moving "vehicles," his emphasis on looks exchanged between strangers points in the right direction. To explore his question of how the experience of mobility in the modern city affects social interaction and collective life, I look more closely at forms of mobile looking.

Mobile Looking

Drawing on work by Harvey Sacks (1989), David Sudnow (1972), and Erving Goffman (1963, 1971), four types of glances or looks used in traffic between unacquainted persons differ in their timing and duration on the one hand, and their focus, on the other: identifying scans, focused looks, sanctioning looks, and integrating glances. This typology will provide a framework for exploring the implications of the speed and material characteristics of mobile assemblages for social interaction.

An identifying scan tells us whether what we are seeing is—in its practical relevancy for our current situated course of action—an expected, "normal," or coherent combination of setting, appearance, and manner, or an incongruent, "abnormal" one (Goffman, 1959, pp. 24-25; Sudnow, 1972, p. 276). Sacks used the example of the incongruence of a "flashy" car driven by someone "who looks like a bum" (Sacks, 1989, pp. 342-343). An identifying scan is brief and unfocused, taking in human and non-human parts of a scene, as "a flick or a shadow of concern" is sufficient to determine whether the observed scene is a "normal appearance" (Goffman, 1971, pp. 238-239). Co-present in spaces of mutual visibility, people make classifications "on the fly" (Joseph, 1998, p. 46) in order to orient their activity by attributing intentions, seeing situations as normal or cause for alarm, perceiving coalitions and collusions, avoiding collisions, and so forth. An incongruent configuration calls for an explanation (Hester & Francis, 2003), and requires a focused look at its incongruous parts.

A focused look interrupts an identifying scan, and settles, however momentarily, on an element—actor, action, or arrangement—in the situation that is noticeably relevant to the observer's current actions (Sudnow, 1972). Goffman (1971) limited these to features of the situation that might be threatening, but there seems no reason to do so; puzzling or aesthetically pleasing features may also attract focused looks.

Because they are, as it were, "one-way," the first two types of looking are not in themselves social interaction, but they are still consequential for mobile action in public places. The production of appearances in public depends on the kind of look which is expected. Where only a momentary look is possible or permissible, actors will tend to act in ways that are interpretable in a glance by a stranger (Sudnow, 1972). An example might be how a pedestrian positions her body so as to signal that she is intending to cross the street, as opposed to waiting on the street corner. Likewise, acts that might appear to the anticipated audience as "occult" or subject to negative attributions will be modified to avoid such interpretations (Goffman, 1963). In the next two types, looks are exchanged and are thus directly a form of social interaction.

A sanctioning look conveys disapproval to its recipient, as when one driver stares disapprovingly at another who has crossed an intersection out of turn. Such a look requires getting the offender's attention in order to deliver the sanction, and can be evaded by the intended recipient knowingly avoiding the other's gaze. As Simmel (1908/1969, p. 148) noted: "shame causes a person to look to the ground to avoid the glance of the other," preventing the sanction from being received. A distinguishing feature of both the sanctioning look and the averted gaze is their temporality, both being held longer than a mere glance. An overly long look can also be aggressive, threatening, or hostile; thus avoiding the appearance of staring as a central component of civil inattention (Goffman, 1963).

Finally, an integrating glance is the mutual glance that "signifies a wholly new and unique union" between persons (Simmel 1908/1969, p. 148), creating "a fantastic kind of social integration" (Sacks, 1989, p. 347). This is the momentary glance in which one person catches another's eye to indicate a shared assessment of the situation. In the example above, the offended driver might exchange looks with another witness to the offence, and without a word being spoken, a temporary bond is formed. This collusion or communion of looks (Goffman, 1959) may be only a momentary form of social solidarity, but when prolonged, it can mobilize the other person for joint action (Goffman, 1971; Joseph, 1998). Integrating glances facilitate coordinated action; by ignoring them, a person can evade cooperative claims (Goffman, 1963).

This typology of mobile looking practices raises two issues that will be the focus of this chapter. First, how is mobile looking by drivers, walkers, or cyclists affected by characteristics of their motion, especially its speed? Second, how is mobile looking affected by material characteristics of their vehicle, especially its shell? Before turning to these issues, Goffman's contribution to a sociological theory of traffic warrants further consideration.

Traffic and the Interaction Order

What kind of social thing is traffic? For there to be "traffic" there must be at least two "vehicular units" that encounter each other and are obliged to take each other into account, even if only to avoid a collision. However much they perform for an absent but potential audience, solitary walkers, cyclists, or drivers are not in or part of traffic. It takes two to traffic.

Goffman's (1963, 1971) analyses of fleeting, mobile encounters between strangers in public make the norms, conventions, and manoeuvres of situated mobile interaction an object of study, and reveal the distinctiveness of its different forms. Automobile, bicycle, and pedestrian traffic can each be considered as both a social order and a diffuse social occasion, composed of unfocused gatherings and unfocused interactions. For Goffman (1963, p. 8):

> a social order may be defined as a consequence of any set of moral norms that regulates the way in which persons pursue objectives. The set of norms does not specify the objectives the participants are to seek, nor the pattern formed by and through the coordination or integration of these ends, but merely the modes of seeking them. Traffic rules and the consequent traffic order provide an obvious example.

Conformity to rules is supplemented with "by-passings, secret deviations, excusable infractions, flagrant violations, and the like" (Goffman, 1971, p. x), so a social order is a product of both the rules—the traffic code—and the manoeuvrings of actors within and beyond its constraints. Much of Goffman's work is thus about the ways in which rules, rituals, and frames are vulnerable to violation, exploitation, or game-playing (Goffman, 1961, 1969, 1983).

In this sense, traffic can be considered a diffuse social occasion in which mobile actors come into each other's presence as part of "a wider social affair, undertaking, or event...[that] provides the structuring social context in which many situations and their gatherings are likely to form, dissolve, and re-form" (Goffman, 1963, p. 18). "Tuesday afternoon rush hour" would be a diffuse social occasion, within which temporary, moving ensembles of pedestrians, cyclists, and automobiles form, gain and lose members, and break up. In such unfocused mobile gatherings, participants manage "sheer and mere copresence" (Goffman, 1963) mainly through identifying scans and the exchange of fleeting focused glances, punctuated by the occasional sanctioning look or integrating glance. As we will see, however, the possibilities for each kind of look, and thus for social interaction is shaped by characteristics of traffic's "vehicular units."

Vehicular Units

Like his metaphors of game, theatre, ritual, and frame (Jacobsen, 2010), traffic played a central role in Goffman's analysis of the interaction order (Quéré, 1989). Nowhere is this more so than in *Relations in Public*, where Goffman (1971) treated pedestrians as "vehicular units," that is, as shells controlled by human pilots or navigators (see Jensen, 2006, pp. 153-154 for a brief discussion). This trick enabled Goffman to roughly sketch some important similarities and differences between automobile traffic and pedestrian traffic in mid-twentieth century American settings.

Goffman (1971, pp. 6-7) identified several similarities between car and foot traffic that also apply to bicycle traffic. They are governed by traffic codes or rules that allow vehicular units independent use of spaces for movement without collision or mutual obstruction. Participants need not be acquainted, so strangers are able to coordinate their passages with some level of mutual trust. As ethnomethodological studies have shown, they are ongoing, collectively produced orders (Hester & Francis, 2003; Livingston, 1987; Ryave & Schenkein, 1974).

Beyond the similarities lie important differences between automobile and pedestrian traffic, according to Goffman (1971, pp. 7-9). Automobile traffic is generally more about getting from point A to point B than is foot traffic, which has more varied purposes. Automobile traffic is typically more linear (Lynch, 1993) than the fluid chaos of pedestrian traffic. The consequences of collision are more serious in automobile traffic than on foot, and the manoeuvres to avoid it are more flexible for pedestrians than for cars (Wolff, 1973), making foot traffic more "permeable" (Demerath & Levinger, 2003). Informal understandings are important in both, but automobile traffic relies more on formal rules than does foot traffic. Automobile traffic is more competitive than pedestrian traffic, in part because drivers can more easily avoid the sanctioning looks of other drivers and more easily escape confrontations. Finally, automobile traffic lacks the "richness of information flow and facilitation of feedback" of face-to-face interaction (Goffman, 1963, p. 17).

Goffman's insights into automobile and pedestrian traffic provide a base for a sociology of traffic. In the remainder of this chapter I extend it by introducing a third type of vehicle and its traffic—bicycles—and by paying more attention than did Goffman to how the speed of vehicular units and the material qualities of their shells afford or constrain interaction with others. This extension will contribute to understanding the differences between walking, driving, and cycling as forms of social life, and also to understanding situational variations within each form of mobility, and in their combinations with other forms of mobility,[1] that will help break down Goffman's categorical distinctions between car and pedestrian traffic.

Pedemobility[2]: The View From the Sidewalk

Walking has recently attracted increasing social scientific interest. Some of it builds on Mauss (1936/1979) by considering walking techniques (e.g., Edensor, 2000, 2010; Ingold, 2004; Vergunst, 2008); other topics include style and identity (Edensor, 2000; Michael, 2000), technologies, especially footwear (Michael, 2000), and embodied experiences (e.g., Edensor, 2008, 2010; Vergunst, 2008). However, little attention has been paid to pedestrian traffic as such. While Goffman has been criticized for emphasizing "visual experience" to the neglect of "experiences of tactile, feet-first, engagement with the world" (Ingold & Vergunst, 2008, p. 3), too much attention to the latter leads to neglect of social interaction on foot, where vision is critical.

If we are to understand pedestrian traffic we need to consider the interaction of unacquainted persons as they encounter each other on paths and sidewalks. The circulation of passers-by is a "succession of thoroughly ritualized arrangements of visibility" (Joseph, 1998, p. 36, my translation). The physical arrangement of bodies and the body language of persons indicate their intentions and relationships, such as "withs" (Jensen, 2010).

The typical sequence of pedestrians passing one another begins with identifying scans, brief glances acknowledging the other and enabling mutual avoidance of collision by reading the other's "routing signals," ending with mutually averted glances to "disattend" the other (Goffman, 1963, p. 84). Considerable efforts can be made to perform identifying scans and focused looks, as Wolff (1973) showed: pedestrians position themselves so as to see beyond the person in front of them, move their heads slightly to monitor whatever is behind them out of the corner of their eyes, and use the expressions of oncoming pedestrians as a rearview mirror to indicate what is going on behind them. A social order arises from conventions and manoeuvres used by pedestrians to avoid "mutual intrusions" on territories of the self, such as not walking beside an unacquainted other and not following someone too closely (Wolff, 1973). Differential speeds force pedestrians (like drivers and cyclists) to focus on "pace management" to avoid such intrusions (Hester & Francis, 2003, p. 43). When they do occur, violations of pedestrian conventions arouse sanctioning looks and hostile comments (Edensor, 2008; Wolff, 1973), perhaps followed by a remedial interchange as accounts and apologies are offered and received (Goffman, 1971).

Observational interchanges can also follow from pedestrian visibility rituals. In the presence of others, a focused look can become a focusing look, as others follow a person's gaze to see what she is looking at (Sacks, 1989). The scene before them can then form a common referent for talk between the observers (Demerath & Levinger 2003). Pedestrian environments often provide sights that provide opportunities for aesthetic pleasures (Lofland, 1998) or "the surprise and pleasure of small, unexpected discoveries" (Lavadinho & Winkin, 2008, p. 161) that are occasions for focused interactions.

Remedial and observational interchanges are made possible by the speed of pedestrians as vehicular units. Speed affects what Demerath and Levinger (2003) called "pausability." The ease of stopping and starting when moving at a walking pace makes it possible to quickly switch from unfocused interactions to focused, face-to-face interactions. The normative constraint of civil inattention limits initiation of interaction, as do mobile involvement shields such as iPods and mobile phones, but the possibility remains.

Yet even at a walking pace, the speed and proximity of fast-moving automobiles and frequent crossings of vehicle spaces (such as driveways, parking lot entrances, and intersections) limits interactional possibilities between pedestrians by forcing them to scan for vehicles rather than other features of the scene. Characteristics of the built environment, such as the width of sidewalks or passageways, may force pedestrian traffic into the more linear pattern characteristic of car traffic, but even in such places, a busker or informal vendor can be accommodated by the manoeuvrability of people on foot.

Despite its neglect of pedestrian traffic, the attention paid to the materiality of technologies and terrain in recent literature on walking raises the question of how the vehicular shell affects possibilities for social interaction. Goffman (1971, p. 7) considered "the individual as pedestrian" as "a pilot encased in a soft and exposing shell" of skin and clothing, but he limited his attention to the expressive, symbolic aspects of the latter. It may be unimportant indoors or in mild weather, but on a cold, windy, winter day, the pedestrian's shell of boots, coat, scarf and hat or hood can restrict all forms of looking. Eyes on the ground, watching for ice or feeling for footing on uneven snow, peripheral vision obscured, does the pedestrian even notice others, and if she does, are they identifiable behind their own warm layers? In other seasons and climates, umbrellas and rain limit what pedestrians can see, although the interactional dance of umbrellas being moved up, down, and sideways to avoid collision with faces or other umbrellas shows that scanning and avoidance of intrusions is usually maintained. Attending to the constraints imposed on pedestrian interaction by the vulnerability of the pedestrian "shell" to its environment would enlarge ethnographic studies of walking. While they do not eliminate the ritual attention paid and received while passing others, material conditions such as terrain, weather conditions, and darkness constrain and modify the identifying scans and focused looks performed by pedestrians. At the same time, because of the pausability of movement on foot, they can also afford topics for conversation and social interaction (Demerath & Levinger, 2003).

The slow speed and the flexible movement of pedestrian vehicular units make possible the features of walking identified by Goffman. Its fluidity, flexibility, and mostly cooperative character, the prevalence of ritual rather than physical damage from collisions, and the dominance of informal negotiated understandings (Wolff, 1973) can occur because glances, and supportive and remedial remarks are easily exchanged between relatively slow-moving, relatively open pedestrian shells. Travel on foot thus provides a baseline for the other forms, whose speed and material shells place more limits on the exchange of glances.

Automobility: The Windshield Perspective

In their pioneering work on automobility, Peter Freund and George Martin (1993, pp. 4–5) referred to the "windshield perspective" of traffic planners and engineers, who historically designed roads with automobiles in mind, treating pedestrians and cyclists as at best an afterthought. The windshield perspective is also an apt term for the visual experience of driving. It is through the windshield (and to a lesser extent the rearview mirror) that drivers engage in unfocused monitoring or identifying scans, and depending on what they reveal, more or less frequent focused looks (Laurier, 2001). As in walking, there is more to the sensory experience of driving than sight, such as the kinaesthetic experience of motion (Sheller, 2004), the sounds of the engine and the tires on the road, and the occasional smell. Nonetheless, driving in traffic is above all a matter of looking and seeing at speed (Edensor, 2003).

The speed of car travel limits opportunities for mutual glances and social interaction. At low speeds or when stopped, eye contact with others is possible, enabling cooperative and ritual courtesies (Goffman, 1967; Jonasson, 1999; Jørgensen, 2008; Vannini, 2011). As speeds rise above 40 or 50 km/h, eye contact becomes less feasible, and the attention of drivers becomes more focused on the road ahead (Jensen, 2006; Laurier, 2001).[3] At higher speeds, social interaction is increasingly unlikely, as "we come close to perceiving the occupants of motor vehicles as not even human, because all we see is a moving object" (Taylor, 2003, p. 1620). The asymmetrical linearity of highway traffic, which is face-to-tail rather than face-to-face, contributes to this (Katz, 1999). In ordinary face-to-face interaction, where sanctioning looks or integrating glances are exchanged, "each giver [of embodied messages] is himself a receiver and each receiver is a giver," as each can see how she is being received by the other, and can be seen to be seeing this (Goffman, 1963, p. 16). In contrast, in ordinary face-to-tail driving, a driver observes vehicles ahead primarily as if they were physical objects (especially because all we usually see of their drivers is the back of their heads, which may even be hidden by headrests). But because they are not just physical objects, we are capable of rapidly shifting into awareness that the vehicle is being piloted by a human, especially whenever our visual scanning detects the absence of normal appearances. At such times movements of other vehicles are morally evaluated as expressions of intention or of involvement (Goffman, 1963) rather than just physical movements, and drivers often become angry at "that idiot" who did such and such deviant act.

The shift from treating other vehicles as physical objects to treating them as expressions of intention occurs abruptly when vehicles rapidly overtake from behind, and the normal scanning of the rearview mirror becomes a focused look. This switch from the forward-looking, windshield perspective to the backward-looking, "rearview mirror perspective" inverts the ordinary exchange of glances of face-to-face interaction. A vehicle in front is seen mainly as a moving physical object, albeit one piloted by a human. A vehicle behind, especially when it is following closely so that the faces of its occupants are visible, is more readily seen as a social object. Thus instead of the situation in face-to-face interaction, where observer and observed are mutually visible as social beings, the face-to-tail linearity of car traffic can create a string of observers who are observed as physical objects by those they are observing as social objects, but are themselves observed as social objects by those they are observing as physical objects.

All this is not to say that social interaction between drivers moving at high speed is impossible, but rather that it is severely constrained. Integrating glances and sanctioning looks are scarcely possible and the visual orientation of drivers is mostly limited to identifying scans and focused looks. The mutual glance, this "most perfect reciprocity in the entire field of human relationships," becomes almost impossible and the situation is dominated by the mere management of co-presence in which "car-drivers are excused from the normal etiquette and social co-ordination of face-to-face interactions" (Sheller & Urry 2000, p. 745).

Unlike walking, when we are driving we are indeed in the situation Simmel (1908/1969, pp. 150-151) described as being "surrounded on all sides by closed doors" and therefore encased in what have been variously called "cyborg bodies" (Lupton, 1999), "car-driver hybrids" (Sheller & Urry 2000; Thrift, 2004), and "car-driver assemblages" (Dant, 2004). Cars, as Urry (2006) has put it, are "inhabited." The automobile shell affects possibilities for looking and social interaction through its role as a backstage (Goffman, 1959) and the effects of height and size. Although "the car functions as a visibility device that makes certain groups recognizable and surveyable to those who are looking in from outside and in particular ways to those gathered together inside" (Laurier et al., 2008, p. 9), occupants tend to treat cars as "invisibility cloaks" by acting in ways which seem to assume that they are not in public (drivers at red lights picking their noses is the classic example). The glass and steel shell of the automobile and the windshield perspective detaches occupants from the exterior scene (unless they are deliberately putting on a performance for an audience outside the vehicle), and they forget that they are in public. Thus Sudnow's (1972) claim that appearances are produced in public knowing that they are observable does not seem to apply to in-car appearances. More than simply relying on the ordinary civil inattention practiced in buses or train cars, occupants of the "living room on wheels" seem to expect the same privacy as in their homes, where passersby are not expected to give more than a passing glance into their windows.[4]

The asymmetry of observing and being observed is complicated by the effect of the differential height and size of vehicles. Larger vehicles block the forward view of smaller vehicles; drivers in higher vehicles can literally look down on others; drivers of heavier vehicles can intimidate smaller ones. This can lead to strategic interaction in which eye contact or its avoidance plays a central role. For example, there are chicken games (Goffman, 1967), as in Kathmandu, where by pointedly not noticing larger vehicles, drivers of smaller ones force the former to yield to them to avoid a collision (Gray, 1994; see also Jørgensen, 2008; Vanderbilt, 2008). Integrating glances facilitate cooperative action; avoiding eye contact is a way of dominating the other in an interaction through what game theory considers the effective, but risky strategy of "asymmetry in communication" (Schelling, 1984, p. 214).

Goffman observed that compared to foot traffic, automobile traffic is more instrumental, linear, competitive, and formally regulated, collisions are more serious, the units in motion less manoeuvrable, and communication between them is more difficult. This section has indicated how these differences depend on the extent to which speed and hard, clumsy shells limit the possibilities for social interaction in automobile traffic.

Velomobility: The View From the Saddle

Like walking, cycling has recently attracted growing interest from social scientists. Unlike walking, that interest has been directed more toward distinctive cycling cultures of mobility and identity (Aldred, 2010; Skinner & Rosen, 2007), especially of bike messengers or couriers (Fincham, 2006, 2007; Kidder, 2009), and to cycling as a form of resistance to automobility (Furness, 2007; Pesses, 2010). There has been little attention to the issues with which this paper is concerned, although the literature contains enough discussion of practices of riding that it is useful for my purposes. In terms of looking, speed, and shells, cyclists are somewhere in between pedestrians and automobiles.

Like drivers, cyclists can move more quickly than pedestrians, creating an obstacle to interaction and the possibility of escape to avoid sanctioning looks and comments in case of transgressions (Jones, 2005). They are generally[5] not as fast as cars however, except in congested traffic, where they may be faster (Kidder, 2009).

Also like automobiles, bicycles are machine-human assemblages, but they lack the protective shell of the former: "the bike is in itself a small object, an object that is embodied and not a space that is inhabited, like a car" (Augé, 2008, pp. 67-68; cf. Aldred, 2010). Consequently, when we are on the saddle, "the external world imposes itself on us concretely in all its most physical dimensions" (Augé, 2008, p. 89), including the muscular effort required to ascend hills that might not even be noticed in an automobile (Wray, 2008) as well as more challenging ascents (Spinney, 2006). Inclement weather, treacherous terrain, and car traffic constrain the cyclist as much as the pedestrian, focusing attention on the road ahead, rather than on interactional possibilities.

Because of the bicycle's small size, lightness, and manoeuvrability, cycling has the pausability and permeability of walking: "the cyclist can stop, chat, and divert from her planned course with relative ease," and cyclists can easily pick up or push their bikes, turning themselves into pedestrians (Aldred, 2010, p. 50; Jones, 2005). They can also form "mobile withs" (Jensen, 2010), such as a couple I saw riding abreast on Boulevard Saint-Michel in Paris one sunny October afternoon, taking up a lane as they chatted.

Speed also makes a difference for what a cyclist sees, and thus for pausability and the exchange of integrating glances. Slower riders on upright city bikes, above the eye level of car drivers and pedestrians, are more open to the sights and sounds around them and to interaction possibilities (Aldred, 2010) than are fast riders of road or racing bikes, who must be focused on scanning the road ahead, especially when riding hard (Spinney, 2006, 2007). But because of their size and manoeuvrability, more speed variation is possible in cycle traffic than in car traffic: fast riders do not tailgate slower ones to get them to speed up; they go around.

As they frequently move in the same spaces as automobiles, and have vulnerable shells, cyclists are even more open to physical and ritual intrusion by automobiles than are pedestrians. In highly automobile-dominated places, such as North America, cyclists are less visible to drivers, who do not expect them, but at the same time their lack of protective shell gives cyclists both the incentive and the ability to be very aware of the vehicles surrounding them. To gain their cooperation, and to ensure that they have been seen, cyclists are often advised to "catch the eye" of drivers. Cyclists (including motorcyclists) are also advised to ride as if they are invisible, that is to not take for granted that drivers will see them. In short, they should ride as if they are not part of normal traffic. But the speed and manoeuvrability of bikes can make automobile traffic permeable for them. When cyclists like the bike messengers studied by Kidder (2009) take advantage of this and do ride as if they were invisible, routinely ignoring red lights, riding against one-way streets, and zipping between moving vehicles, they become all too visible to drivers, evoking hostility and anger because they contradict "normal appearances" and their intentions are not apparent. Although not threatening in themselves, opaque intentions may "leave the witness not knowing where the mind of the performer is, or what his purpose, and therefore not trustful of him" (Goffman, 1971, p. 306). (Even though I cycle, as a driver I often find the behavior of some cyclists to be unpredictable and inexplicable).

Figure 1. The windshield perspective, the rearview mirror perspective, and cyclist visibility. Billboard, London, England, November 2010. Photo by the author.

Although "fundamental differences between the needs and experiences of walkers and cyclists" have been claimed (Middleton, 2011, p. 92), I have tried to show here that the differences are a matter of degree, the shells and speed of bicycles putting them somewhere between automobile and pedestrian traffic.

Conclusion

Despite the "mixed status" of its evidence—a fault shared with Goffman's work (1959, p. xi) without the compensation of his style and originality—I hope that the argument presented in this paper can convince the reader that there are benefits to closely examining the conditions enabling and constraining practices of looking in traffic. Starting from Simmel's remarks on the sociological significance of mutual glances, a typology of looks was developed from Goffman and from ethnomethodologists such as Sacks and Sudnow. Goffman's analysis of social orders and diffuse social occasions was then used to further understand the interactional role of looking for vehicular units in traffic. By remedying Goffman's neglect of two qualities of these vehicular units—their shells and speed of movement—I have been able to sketch some of the significant differences between pedestrian, bicycle, and automobile traffic. In comparing these types, much subtlety has been sacrificed, but I hope that enough has been done to show that such comparisons are productive. Comparisons involving mobilities which, like velomobility, are in between automobility and pedemobility should also be productive. For example, studies of the mobilities of rollerblades, skateboards, motorized wheelchairs, and other hard-to-classify technologies (Cox, 2007) will further help in understanding limits and possibilities of co-presence and interaction in urban spaces.

In making such comparisons between mobilities, we must be careful not to ignore different cultures of mobility in different places and situations, in order to avoid over-generalizing. For example, Wolff's (1973) observation, made in Manhattan in the third quarter of the twentieth century, of children being dragged along by their parents like baggage is reported by Ingold (2004, p. 328) as if true of "city parents" in general. My observations on streets in Paris in the autumn of 2010 contradict this: much like Batek hunter-gatherers in Malaysia (Ingold & Vergunst, 2008), in the daytime at least, parents there often trailed behind their children or allowed their children to trail behind them.

Cultural and situational variations raise more than just methodological issues. They also raise theoretical ones concerning the relative importance of dispositional and situational explanations of action. On the one hand, it appears that people are capable of switching quickly from one form of mobility to another and from one mode of interaction to another. For example, drivers getting out of their cars are immediately capable of interacting like pedestrians, and cyclists can become pedestrians by picking up their bikes. Switching from one means of mobility to another is a switch into a different situation, with different traffic codes and social conventions, and different material conditions for the exercise of vision, speech, and other senses (nicely illustrated in Vannini & Vannini, 2008). Thus it is not habitus but the situation, and both its social or ritual conventions and its physical conditions—speed and shells—that explain interactions.

Switching is important in another sense. The metaphors of ritual, theatre, game, and frame that Goffman used are not just metaphors, they are ever-present possibilities in all forms of human traffic. Pedestrians, cyclists, and drivers can and do switch from exchanging ritual care for each other's sacred selves, to performing for a real or imagined audience, engaging in contests for fun or profit, or exploiting the myriad vulnerabilities and layerings of frames (Goffman, 1974). Exploring how traffic codes and the speeds and shells of the interacting vehicular units affect these possibilities promises theoretical benefits.

On the other hand, dispositions or habitus are often seen as determining action. Does the predominance of one form of mobility lead its practitioners to extend its modes of interaction to another? Does the tendency of male cyclists to ride as fast as they can (a practice of which the author of this paper is guilty) dispose them to drive as fast as they can (the author pleads innocent in this case). Do the automobile identities of "fast driver" and "slow driver" (Sacks, 1992), created in the linearity and instrumentality of driving, leak into pedestrian behavior? Reports of "pedestrian rage" between "fast walkers" and "slow walkers" suggest that they have, although the existence of an "I hate slow walkers" Facebook group, or the comments of anonymous posters on a newspaper website who assert that foot traffic should follow the same rules as car traffic only hint at what might be involved (Anderssen, 2011).

Use of Goffman and ethnomethodology in this paper indicates where the author's theoretical dispositions lie, but the debate is not over. Although they may not have the cachet of aeromobilities or current forms of mediated electronic and virtual mobility, foot, bike, and car traffic have much to teach us still.

Acknowledgment

For helpful comments on previous versions of this chapter, I thank Pradeep Bandyopadhyay, Jim Cosgrave, Arlene McLaren, Phillip Vannini, and the anonymous reviewer for this publication. As usual, they are not responsible for my errors and infelicities, especially as I did not always take their advice.

Notes

1. Because of space limitations, the combination of pedestrian, driver, and cyclist mobilities will only be partially explored here.
2. "Vélomobility" is used by Furness (2007) and Pesses (2010). Like "aeromobility," it does not otherwise seem to exist in English, so at the risk of further degrading the language I add the neologism, "pedomobility," for the sake of symmetry. "Pedomobility" sounds better, but use of the term "pedomobile" as slang for a car driven by a pedophile rules it out.
3. Above 30 km/h, collisions become lethal for pedestrians (Vanderbilt, 2008, pp. 194-195); 50 km/h is the usual speed limit on city streets in Canada. A colleague who was first learning how to drive in her 50s told me it was difficult to unlearn habits acquired as a passenger, such as looking all around, instead of scanning the road ahead.
4. This paragraph is indebted to past students in my sociology of the automobile class at Trent University: one who called cars invisibility cloaks; another who (unbeknownst to me) had a friend drive him on a local freeway, where he peered into the interiors of cars as they drew alongside, to see what occupants were doing. For this breach of privacy (and research ethics), he reported frequently being given "the finger."
5. The experience of driving down a steep, long, and winding hill in Normandy, as fast as I considered safe—about 80-90 km/h—with two cyclists right behind me all the way is why I add the qualifier "generally."

References

Aldred, R. (2010). 'On the outside': Constructing cycling citizenship. *Social and Cultural Geography*, 11(1), 35-52.
Anderssen, E. (2011, February 15). *Why do slow walkers enrage some people and not others?* Globe and Mail Blog. Retrieved from http://www.theglobeandmail.com/life/the-hot-button/why-do-slow-walkers-enrage-some-people-and-not-others/article1908209/
Augé, M. (2008). Éloge de la bicyclette. Paris, France: Éditions Payot & Rivages.

Cox, P., with Van De Walle, F. (2007). Bicycles don't evolve: Velomobiles and the modelling of transport technologies. In D. Horton, P. Rosen, & P. Cox (Eds.), *Cycling and society* (pp. 113-131). Farnham, UK: Ashgate.

Dant, T. (2004). The driver-car. *Theory, Culture & Society*, 21(4-5), 61-79.

Demerath, L., & Levinger, D. (2003). The social qualities of being on foot: A theoretical analysis of pedestrian activity. *City and Community*, 2(3), 217-237.

Edensor, T. (2000). Walking in the British countryside: Reflexivity, embodied practices and ways to escape. *Body & Society*, 6(3-4), 81-106.

Edensor, T. (2003). Defamiliarizing the mundane roadscape. *Space and Culture*, 6(2), 151-168.

Edensor, T. (2008). Walking through ruins. In T. Ingold & J. L. Vergunst (Eds.), *Ways of walking: Ethnography and practice on foot* (pp. 123-142). Aldershot, UK: Ashgate.

Edensor, T. (2010). Walking in rhythms: Place, regulation, style and the flow of experience. *Visual Studies*, 25(1), 69-79.

Fincham, B. (2006). Bicycle messengers and the road to freedom. *Sociological Review*, 54(1), 208-222.

Fincham, B. (2007). Bicycle messengers: Image, identity and community. In D. Horton, P. Rosen, & P. Cox (Eds.), *Cycling and society* (pp. 179-195). Farnham, UK: Ashgate.

Freund, P., & Martin, G. (1993). *The ecology of the automobile*. Montreal, Canada: Black Rose Books.

Furness, Z. (2007). Critical mass, urban space and vélomobility. *Mobilities*, 2(2), 299-319.

Goffman, E. (1959). *The presentation of self in everyday life*. Garden City, NY: Doubleday.

Goffman, E. (1961). *Encounters: Two studies in the sociology of interaction*. Indianapolis, IN: Bobbs-Merrill.

Goffman, E. (1963). *Behavior in public places: Notes on the social organization of gatherings*. New York, NY: Free Press.

Goffman, E. (1967). *Interaction ritual: Essays on face-to-face behavior*. Garden City, NY: Anchor Books.

Goffman, E. (1969). *Strategic interaction*. Philadelphia, PA: University of Pennsylvania Press.

Goffman, E. (1971). *Relations in public: Microstudies of the public order*. New York, NY: Harper & Row.

Goffman, E. (1974). *Frame analysis: An essay on the organization of experience*. New York, NY: Harper & Row.

Goffman, E. (1983). The interaction order. *American Sociological Review*, 48(1), 1-17.

Gray, J. (1994). Driving in a soft city: Trafficking in images of identity and power on the roads of Kathmandu. In M. Allen (Ed.), *Anthropology of Nepal: Peoples, problems and processes* (pp. 147-159). Kathmandu, Nepal: Mandala Book Point.

Hester, S., & Francis, D. (2003). Analysing visually available mundane order: A walk to the supermarket. *Visual Studies*, 18(1), 36-46.

Ingold, T., (2004). Culture on the ground: The world perceived through the feet. *Journal of Material Culture*, 9(3), 315-340.

Ingold, T. & Vergunst, J. L. (2008). Introduction. In T. Ingold & J. L. Vergunst (Eds.), *Ways of walking: Ethnography and practice on foot* (pp. 1-19). Aldershot, UK: Ashgate.

Jacobsen, M. H. (2010). Introduction: Goffman through the looking glass: From 'classical' to contemporary Goffman. In M. H. Jacobsen (Ed.), *The contemporary Goffman* (pp. 1-47). New York, NY: Routledge.

Jensen, O. B. (2006). 'Facework', flow and the city: Simmel, Goffman, and mobility in the contemporary city. *Mobilities*, 1(2), 143-165.

Jensen, O. B. (2010). Erving Goffman and everyday life mobility. In M. H. Jacobsen (Ed.), *The contemporary Goffman* (pp. 333-351). New York, NY: Routledge.

Jonasson, M. (1999). The ritual of courtesy—Creating complex or unequivocal places? *Transport Policy*, 6(1), 47-55.

Jones, P. (2005). Performing the city: A body and a bicycle take on Birmingham, UK. *Social and Cultural Geography*, 6(6), 813-830.

Jørgensen, A. J. (2008). The culture of automobility: How interacting drivers relate to legal standards and to each other in traffic. In T. P. Uteng & T. Cresswell (Eds.), *Gendered mobilities* (pp. 99-111). Aldershot, UK: Ashgate.

Joseph, I. (1998). *Erving Goffman et la microsociologie*. Paris, France: Presses universitaires de France.

Katz, J. (1999). *How emotions work*. Chicago, IL: University of Chicago Press.

Kidder, J. L. (2009). Mobility as strategy, mobility as tactic: Post-industrialism and bike messengers. In P. Vannini (Ed.), *The cultures of alternative mobilities: Routes less travelled* (pp. 177-191). Aldershot, UK: Ashgate.

Laurier, E. (2001). Notes on dividing the attention of a car driver. *TeamEthno-Online Journal 1*. http://www.teamethno-online.org.uk/Issue1/Laurier/ gooddriv.html.

Laurier, E., Strebel, I., Lorimer, H., Brown, B., Jones, O., Juhlin, O., et al. (2008). Driving and 'passengering': Notes on the ordinary organization of car travel. *Mobilities*, 3(1), 1-23.

Lavadinho, S., & Winkin, Y. (2008). Enchantment engineering and pedestrian empowerment: The Geneva case. In T. Ingold & J. L. Vergunst (Eds.), *Ways of walking: Ethnography and practice on foot* (pp. 155-168). Aldershot, UK: Ashgate.

Levine, J., Vinson, A., & Wood, D. (1973). Subway behavior. In A. Birenbaum & E. Sagarin (Eds.), *People in places: The sociology of the familiar* (pp. 208-216). New York, NY: Praeger.

Livingston, E. (1987). *Making sense of ethnomethodology*. London, UK: Routledge & Kegan Paul.

Lofland, L. H. (1998). *The public realm: Exploring the city's quintessential social territory*. Hawthorne, NY: Aldine de Gruyter.

Lupton, D. (1999). Monsters in metal cocoons: 'Road rage' and cyborg bodies. *Body & Society* 5(1), 57-72.

Lynch, M. (1993). *Scientific practice and ordinary action: Ethnomethodology and social studies of science*. New York, NY: Cambridge University Press.

Mauss, M. (1979). Body techniques. In M. Mauss, *Sociology and psychology: Essays* (B. Brewster, Trans., pp. 95-123). London, UK: Routledge and K. Paul.

Michael, M. (2000). These boots are made for walking...: Mundane technology, the body and human-environment relations. *Body & Society*, 6(3-4), 107-126.

Middleton, J. (2011). Walking in the city: The geographies of everyday pedestrian practices. *Geography Compass*, 5(2), 90-105.

Pesses, M. W. (2010). Automobility, vélomobility, American mobility: An exploration of the bicycle tour. *Mobilities*, 5(1), 1-24.

Quéré, L. (1989). 'La vie sociale est une scène': Goffman revu et corrigé par Garfinkel. In I. Joseph, R. Castel, J. Cosnier, et al., *Le parler frais d'Erving Goffman: Colloque de Cerisy* (pp. 47-82). Paris, France: Editions de Minuit.

Ryave, A. L., & Schenkein J. N. (1974). Notes on the art of walking. In R. Turner (Ed.), *Ethnomethodology* (pp. 265-274). Harmondsworth, UK: Penguin.

Sacks, H. (1989). Lecture eleven: On exchanging glances. *Human Studies*, 12(3-4), 333-350.

Sacks, H. (1992). Lecture 24: Systems of measurement. In H. Sacks, *Lectures on conversation*, vol. 1 (pp. 435-440). Oxford, UK: Blackwell Publishing.

Schelling, T. C. (1984). What is game theory? In T. C. Schelling, *Choice and consequence: Perspectives of an errant economist* (pp. 213-242). Cambridge, MA: Harvard University Press.

Schmidt-Relenberg, N. (1986). On the sociology of car traffic in towns. In E. de Boer (Ed.), *Transport sociology: Social aspects of transport planning* (pp. 121-132). Oxford, UK: Pergamon Press.

Sheller, M. (2004). Automotive emotions: Feeling the car. *Theory, Culture & Society*, 21(4-5), 221-242.

Sheller, M., & Urry, J. (2000). The city and the car. *International Journal of Urban and Regional Research*, 24(4), 737-757.

Simmel, G. (1969). Sociology of the senses: Visual interaction. In R. E. Park & E. W. Burgess (Eds.), *Introduction to the science of sociology, student edition* (pp. 146-151). Chicago, IL: University of Chicago Press.

Simmel, G. (1971a). The metropolis and mental life. In D. N. Levine (Ed.), *Georg Simmel on individuality and social forms: Selected writings* (pp. 324-339). Chicago, IL: University of Chicago Press.

Simmel, G. (1971b). The stranger. In D. N. Levine (Ed.), *Georg Simmel on individuality and social forms: Selected writings* (pp. 143-149). Chicago, IL: University of Chicago Press.

Skinner, D., & Rosen, P. (2007). Hell is other cyclists: Rethinking transport and identity. In D. Horton, P. Rosen, & P. Cox (Eds.), *Cycling and society* (pp. 83-96). Farnham, UK: Ashgate.

Spinney, J. (2006). A place of sense: A kinaesthetic ethnography of cyclists on Mont Ventoux. *Environment and Planning D: Society and Space*, 24(5), 709-732.

Spinney, J. (2007). Cycling the city: Non-place and the sensory construction of meaning in a mobile practice. In D. Horton, P. Rosen, & P. Cox (Eds.), *Cycling and society* (pp. 25-45). Farnham, UK: Ashgate.

Sudnow, D. (1972). Temporal parameters of interpersonal observation. In D. Sudnow (Ed.), *Studies in social interaction* (pp. 259-279). New York, NY: Free Press.

Taylor, N. (2003). The aesthetic experience of traffic in the modern city. *Urban Studies*, 40(8), 1609-1625.

Thrift, N. (2004). Driving in the city. *Theory, Culture & Society*, 21(4-5), 41-59.

Urry, J. (2004). The 'system' of automobility. *Theory, Culture & Society*, 21(4-5), 25-39.

Urry, J. (2006). Inhabiting the car. *Sociological Review*, 54(1), 17-31.

Urry, J. (2007). *Mobilities*. Oxford, UK: Polity Press.

Vanderbilt, T. (2008). *Traffic: Why we drive the way we do (and what it says about us)*. New York, NY: Alfred A. Knopf.

Vannini, P. (2011). Mind the gap: The tempo rubato of dwelling in lineups. *Mobilities*, 6(2), 273-299.

Vannini, P., & Vannini, A. (2008). Of walking shoes, boats, golf carts, bicycles, and a slow technoculture: A technography of movement and embodied media on Protection Island. *Qualitative Inquiry*, 14(7), 1272-1301.

Vergunst, J. L. (2008). Taking a trip and taking care in everyday life. In T. Ingold & J. L. Vergunst (Eds.), *Ways of walking: Ethnography and practice on foot* (pp. 105-122). Aldershot, UK: Ashgate.

Wolff, M. (1973). Notes on the behavior of pedestrians. In A. Birenbaum & E. Sagarin (Eds.), *People in places: The sociology of the familiar* (pp. 35-48). New York, NY: Praeger.

Wray, J. H. (2008). *Pedal power: The quiet rise of the bicycle in American public life*. Boulder, CO: Paradigm Publishers.

13

Imaginative Technologies of (Im)mobility at the "End of the World"

Noel B. Salazar

During the southern hemisphere summer of 2009-2010, I was in Chile conducting research on transnational mobilities. The country had just been admitted as the first South American member of the OECD, the so-called "rich man's club" of nations and, capitalizing on this, the government was busy preparing its *Chile hace bien* ("Chile is good for you") campaign to promote Chile abroad. I was interested in the (dis)connections that Chileans make between becoming a "developed" country and (increased) transnational mobility. Interestingly, the people I talked to focused mostly on explaining why they personally wanted to stay put. Comments such as: "I wouldn't leave, in my case I'm rooted in my family and I'm not desperate to leave the country" were common. People also stated that "Chile is so far from everything." While many Chilean citizens do not have the financial means to travel, increased numbers of academic scholarships and overseas work placements are providing the opportunity to study or work abroad. Nevertheless, many Chileans stress that they are homebound and draw on a metaphor of their country as an inaccessible island, with both its positive quality of insulation and its negative characteristic of isolation, as justification (cf. Vannini, 2011). Much of this imaginary derives from the country's history. This chapter reports on what I discovered and suggests how it helps us to understand contemporary (im)mobilities in Chile and beyond.

This edited volume deals in particular with technologies of mobility. Here, technology is defined as any technique, system, or method of organization that serves a particular purpose (mostly related to human ability to control and adapt to varying environments). Technologies of (im)mobility, then, are utilized to control and adapt to various forms of sedentarism and movement. As the other contributions to this book richly illustrate, technologies of mobility encompass a wide range of practices and procedures. I argue in this chapter that, apart from "spatial, infrastructural and institutional moorings" (Hannam, Sheller, & Urry, 2006, p. 3), there is another type of technology that configures and enables or constrains boundary-crossing mobilities. The technology I am disentangling here may be less visible than other categories, but is certainly as important (Brann, 1991; Strauss, 2006). Historically laden imaginaries—culturally shared and socially transmitted representational assemblages that are used as meaning-making devices—are the "energetic source" (Baeza, 2008, p. 24) that inspires social life, including people's (im)mobilities (Salazar, 2010a, 2010b). Imaginaries of mobility, or "representations of movement that give it shared meaning" (Cresswell, 2010, p. 19), form an important part of the ideas, beliefs, or habits to which people are accustomed and from which they gain security or stability in both their daily activities and the long-term planning of their lives.

Imaginative technologies, cultural mechanisms enacted through socially shared imaginaries, create subjectivities which are used by individuals as personal resources for the construction and management of their identities and activities. People worldwide rely on unspoken, powerful images and ideas, from the most spectacular fantasies to the most mundane reveries, to shape multiple, often conflicting, identifications of Self and Other. Imaginaries of the world (including the Other) and of oneself, as individual or as part of society, always go hand in hand (Castoriadis, 1987). These constructions, whether descriptive or normative, may maintain a subtle distribution of power and privilege, which may not be obvious to the people who have internalized them (Jónsson, 2011). While social imaginaries are part of the glue that holds groups together (Taylor, 2004), one of the central problems is the lack of correspondence between the projected ideals and aspirations on the one hand, and the perceived and experienced reality on the other. No wonder utopias are one of the common expressions of the social imaginary. These are critical visions of good and possibly attainable social systems and lives, either spatial (located elsewhere) or temporal (located in another time).

Boundary-crossing mobilities, too, involve how people form relations with others and how they make sense of this (Adey, 2006). Their meaning and experience is always tied to certain dominant, sociocultural values and expectations. People hardly journey to *terrae incognitae*, but to destinations they already virtually "know" through the widely circulating imaginaries about them. Such imaginaries travel through a multitude of channels and provide the cultural material to be drawn upon and used for the creation of translocal connections. Empowered by mass-mediated images and discourses, these culturally inflected imaginaries have changed the way in which people collectively envision the world and their own position and mobilities within it (Salazar, 2010a). Imaginaries of mobility play a constitutive role in social structuration and can be seen as pervading "constellations of mobility," which Cresswell defined as "historically and geographically specific formations of movements, narratives about mobility and mobile practices" (2010, p. 17). These constellations are "multi-faceted, diverse, never subject to simple characterizations" (Vannini, 2011, p. 267).

This chapter develops Cresswell's approach to take "both historical mobilities and forms of immobility seriously" (2010, p. 17). Using Chile as a case study, I explore how the dominant imaginaries circulating about this country, both inside and outside its borders, are "foreign." Their cyclical repetition throughout Chile's history shows their effectiveness as technologies of (im)mobility. Fieldwork was conducted in Chile between December 2009 and January 2010. Methods included archival research, observation (direct or participant) and free-flowing interviews with key informants and other significant actors in the field of transnational mobility (mainly migration and tourism). Ancillary data included audio-visual material, news media, documents, and websites. Findings were recorded in personal research diaries. In this chapter, I sketch the historical genealogy of imaginaries about Chile, many of which are utopian in nature, and then focus on how old imaginaries of Chile as "the end of the world" impact on how contemporary Chileans participate in, and frame their perceived exclusion from, a plethora of new mobilities, regardless of whether they have the means and freedom to cross (imaginary and real) boundaries.

The End of the World (as People Know It)

It is a bright Saturday morning and I am in downtown Santiago, Chile's bustling capital. While I await the arrival of one of my informants, I have time to observe the surroundings. It is hard to ignore the multiple markers showcasing the city's global interconnectedness: global brand name stores and restaurant chains, advertisements for European and North American movies, the occasional foreign face in the crowd, and Chileans on cell phones referring to faraway worlds. This observation stands in marked contrast with a nationwide research study conducted by the United Nations Development Program which revealed that 38% of young people in Chile are not receptive to "foreignness" and have never considered living abroad (PNUD, 2003, pp. 13-14). Such attitudes are particularly pronounced amongst girls and people belonging to the middle or lower social classes, while boys from middle and higher classes tend to hold opposing views and are much more receptive to foreign influences (PNUD, 2003, pp. 14-15). Young people living in the capital were found to display the most cosmopolitan attitudes.

My informant is a young history student from a wealthy Chilean family. After only a few minutes' discussion, it becomes clear that Juan (pseudonym) holds serious reservations about transnational mobility. He has many opportunities to travel abroad but never makes concrete plans. He likes to fantasize about the possibility, but seems afraid to realize a border-crossing trip (and to leave the safety of "the island"). Such mobility-related anxieties have been documented elsewhere, too (Lindquist, 2009). Apart from mentioning strong family bonds, another reason Juan gives me is that he is afraid that "seeing Chile from the outside can be a disenchanting experience" (referring to the fact that there is a danger of not recognizing one's own lifeworld or "different-because-remote" frame of reference anymore if one stays abroad for too long). Even if he acknowledges how valuable an experience travelling abroad can be, out of compliance, Juan, in common with many of my other research participants, prefers to stay in Chile. When I ask Juan why so few Chileans want to study or work abroad, he suggests to me that having a closer look at the country's history may help to understand contemporary Chilean society's attitude towards transnational mobilities.

Chile is a 4,300 km long and narrow strip of land (on average 175 km wide) between the Andes Mountains in the east and the Pacific Ocean in the west, the Atacama Desert in the north and the icebergs of Patagonia in the south. This "crazy geography" (Benjamín Subercaseaux, 1973) has not only determined its territorial boundaries, but also influenced the imaginaries both natives and foreigners alike have about the country. Prior to the arrival of Spanish colonizers in the sixteenth century, northern Chile was under Inca rule and various groups of Araucanian Indians inhabited the central lands and southern islands. The origin of the word *Chile* is contested. One possible linguistic genealogy is the Aymara concept *chilli* ("where the world ends"). For the Aymara, a native ethnic group living in the Andes and the Altiplano, it made sense to denote the lands southwest of theirs as the end of their lifeworld. The Spanish had done the same at home, calling the westernmost point of the Iberian Peninsula Cabo Fisterra (Cape Finisterre). However, these geographical imaginaries drastically changed when Iberians discovered, in the fifteenth century, that, far beyond their western "edge of the world," there was another land: America. From that moment onwards, the so-called New World became one of the favorite places onto which Europeans projected their wildest images and ideas of paradise on earth and where many of the old continent's failed utopias could materialize.

Most cultures and religions have myths depicting an imaginary existence different from the hardships of real life, an existence blessed with nature's bounty, untroubled by strife or want. This happy state is nearly always placed somewhere or sometime outside normal human experience, often "off the map" in some remote quarter of the world. Such ideas from the Old Testament and ancient Greco-Roman and Celtic myths played an important role in providing some of the explorers and early settlers in the New World with a framework to understand, explain, and justify their activities (Aínsa, 1999; Baritz, 1961). In the words of Lévi-Strauss, "when they moved into unknown regions they were more anxious to verify the ancient history of the Old World than to discover a new one" (1955/1961, p. 78). As the discoverers of the late fifteenth and early sixteenth centuries came to venture to the edges of the world, it was supposed that, sooner or later, they would encounter some of the mythical geographical utopias and figures whose existence was, at least for a great number of them, beyond dispute.

In the age of the discoveries, the whole of the Americas fulfilled the legendary role of the "end of the world." Yet, the utopian America that had to be was systematically crushed under the real one. With the way the conquests went, North and Central America gradually lost their mythical qualities and the European imaginary of the end of the world moved from the "Far West" to the "Far South" (Franz, 2000). As Stuven wrote, "for the Spanish Empire, Chile was one of the least important colonies; it was extremely far from the centers of power and it was a region of intermittent warfare. Who wanted to go to Chile? Very few. And this inevitably influenced and still influences the Chilean character" (2007, p. 47).[1] Stuven, a Chilean historian, gave a good example of the long-lasting influence of deep-rooted foreign imaginaries. Chile was, indeed, a poor colony; the colonizers never found the extensive gold and silver they had anticipated (partially based on the legends of King Solomon's mines and El Dorado). On the contrary, the conquest of Chile probably cost Spain more blood and treasure than all the rest of America. Interestingly, Chile also became one of Europe's ultimate utopian playgrounds, both in fiction (for writers and armchair adventurers) and in reality. Many wrote convincingly about Chile without ever having been there themselves (see Roa & Teillier, 1994).

By calling Chile "a fertile province" in his poem *La Araucana* (1589), the Spanish soldier, Alonso de Ercilla, gave rise to the myth of Chilean exceptionalism. Although Ercilla himself was barely a year and a half in Chile, his epic poem that sang the Spanish Empire of Philip II, already sensed Chile as a nation. No wonder Pablo Neruda, whose own writings were inspired by the work of Ercilla, called the latter the "inventor of Chile." The ways in which such discursive technologies play a constitutive role in social structuration has been well documented (Blommaert & Bulcaen, 2000). The social-historical imaginaries of Chile as an extraordinary country, be it for its natural beauty, its political achievements, its culture, its military victories, its resilience in time of adversity, and numerous other qualities, have deeply influenced intellectuals and artists alike (cf. Stuven, 2007, p. 55).

Once independent, liberal Chilean politicians fought the colonial stigma that defined their country as the poorest and most miserable region of the New World, turning this imaginary around by means of discourses of geographical uniqueness to make Chile a "happy copy of Eden" (as evoked in the national anthem and recycled in popular culture elements). They used a related set of discursive technologies to redefine the country in the political sphere as the antithesis of South American reality, thanks to the institutional stability of the ruling elites. This is a nice example of what Cresswell (2010, p. 21) referred to as "politics of representation," which impact on the experience of (im)mobilities. While the name of Chile in the colonial era had been associated with loss, isolation, violence, and insecurity, after independence it represented not only the republican ideal, but also stability and order (Sagredo Baeza, 2006). In other words, the natural condition and the geographic location of Chile conditioned not only its colonial image and development, but also its organization as a Republic. In 1830, the French naturalist, Claude Gay, was contracted to travel throughout Chile for three and a half years, investigating everything from geography and geology to demographics and industry. His monumental *Atlas de la historia física y política de Chile* (published in 30 volumes between 1844 and 1871) not only mapped out the country's political and natural past, but also greatly contributed to the Chilean cultural identity (Mizón, 2001).

Many of the foreign descriptions of Chile after the country had become independent refer to the extreme south, for the simple reason that, before the opening of the Panama Canal in 1914, the Strait of Magellan was the main route for steam ships travelling from the Atlantic Ocean to the Pacific (Canihuante, 2006, p. 86). The ruling elite used this focus on the "cold," southern part to distinguish the country from its Latin American neighbors. Following the lines of eighteenth- and nineteenth-century European discourses linking climate to racial development, Chilean authors insisted upon the tropical character of neighboring countries in contrast with the (European-like) temperate climes of Chile. This kind of thinking was backed up by the argument that the word *Chile* actually goes back to the Quechua word *chiri* ("cold") or the Aymara word *ch'iwi* ("shadow"), which is, coincidentally, pronounced the same as the English *chilly* (Pérez de Arce, 2006, p. 19).

A Short History of Migratory Mobilities

What are the (dis)connections between imaginaries, such as the ones described above, and people's mobilities? The history of transnational mobility in Chile starts with the Spanish conquistadors. Although they were among the first international immigrants that arrived in the country, they are not often recognized and named as such (Cano & Soffia, 2009). While colonizers also carry with them the flag of their country of origin to try to impose it on foreign territory, migrants mostly seek utopia in supposedly ideal places (the imagined

Promised Land), escaping from realities of submission and misery, if not persecution, that push them to leave their native place without titles or belongings (Aínsa, 1999). During the sixteenth and seventeenth centuries, Chile received many Spanish migrants, mainly from Extremadura and Castile and León, together with a small group of slaves of African descent. In the eighteenth century, people from Basque Country arrived, together with British and French traders. Most scholars, however, only started using the concept of migration when Chile gained independence and welcomed European soldiers and maritime traders, particularly English, French, and Italian people (but also Dutch, Greek, Portuguese, and Scandinavian), which in turn facilitated the spontaneous arrival of other Europeans. Many European governments sponsored migration to Chile.

The waves of immigrants who left Europe between 1850 and 1914 helped to reinvigorate the foundational spirit of the Promised Land; Latin America was seen to be the space and time of utopia. Nevertheless, Chile developed as a socially and culturally insular country unaccustomed to the presence of large numbers of foreigners. The early Chilean governments had two main motives for attracting European migrants: the colonization of the south of the country (finishing the task that the Spanish colonizers had begun), and the widespread belief that the Europeans, as hard workers, would foster economic development and modernization. Although the overall number of immigrants during this early period was relatively small, their presence transformed the country technologically, economically, religiously, and culturally. Already in 1824, the government enacted a law to encourage Europeans (primarily Swiss, German, and English) to establish factories in urban centers as well as to populate sparsely inhabited southern areas.

The first admission of immigrants to Chile was selective. In 1845, an immigration law ("Ley de Colonización") prescribed how the migration process was to unfold (Zavala San Martín & Rojas Venegas, 2005). The 1854 census shows approximately 20,000 foreigners, most of them German colonists in the Region of the Lakes (predominantly village artisans and agriculturalists). Their settlements in the south of Chile are a good example of how agricultural colonies were formed that transpose religions, customs, and architecture. As Baeza (2008) noted, "when these settlers wrote to their loved ones who had remained in Germany, they often evoked the illusion of a promised land, relying on a pan-Germanic ideal in which southern Chile was merely inserted as an extension of the Black Forest that, of course, the weather conditions and southern flora made them recall with nostalgia" (p. 265). In 1882, the immigration effort was reinforced through the establishment of the country's *Agencia General de Colonización* (General Immigration Agency) in European ports, offering Chilean land in uncultivated areas to settler families. Despite these efforts, relatively few immigrants actually came—between 1889 and 1907, for instance, only 55,000 arrived, while Argentina in the same period received well over two million. Those who did arrive increasingly came on their own account, and not, as occasionally in the past, as part of government-sponsored immigration schemes.

The years between the 1907 and 1952 censuses are notable for the growth of immigrant populations of Arabs (fleeing conflicts in the Ottoman Empire, which then included Syria, Palestine, and Lebanon) and Asians. This migration was largely undocumented. Because these migrants were not White, they were not welcomed as warmly as their European predecessors, as they were considered to be both culturally inferior and an economic threat to the country (see Collier & Sater, 2004; Doña & Levinson, 2004).

The industrialization of the 1940s and the end of World War II prompted Chile to look for specific types of migrants. The 1952 migration law is very explicit about which kind of migrants were needed to contribute to the country's new economy. Internally, the country went through a rapid succession of divergent political ideologies in the second half of the twentieth century. From 1964 to 1970, there was a "revolution in liberty" (a social-Christian utopia) under Eduardo Frei; from 1970 a "Chilean road to socialism" (a Marxist utopia) under Salvador Allende; and from 1973 to 1989 a "silent revolution" (a neo-liberal, capitalist utopia) under dictator, Augusto Pinochet. The decree-law of 1975 detailed four desired groups of foreigners: students, tourists, workers, and residents. Paradoxically, this type of policy served to both diversify the foreigners in Chile and to restrict access for political and ideological reasons. More importantly, the dramatic political changes turned Chile from an immigrant to an emigrant nation (Martínez Pizarro, 2003). Pinochet's new social, political, and economic order undoubtably discouraged many potential immigrants. It is estimated that during the period 1973-1985 between 500,000 and 1,000,000 Chileans left their country, either voluntarily or compulsorily (Jedlicki, 2001). At the same time, the military dictatorship used the argument of distance to dismiss all foreign criticism: one should not listen to what is being said about Chile abroad; nobody understands us because we are different (Pizarro, 2003, p. 106).

Chile, a "Cool" Place

After having endured a difficult, 17-year dictatorial period (1973-1990), the military regime had isolated Chile from the world (and, thus, from transnational migrants and tourists). The only sector of Chilean society that was successful in creating a positive image abroad was the business sector, which could boast the successes of the "Chilean miracle" (having turned Chile into one of the most prosperous nations of South America). The way in which the Chilean state took it upon itself, after the return to democracy, to "reinsert" Chile into the international community was partly influenced by the rising migration from within South America, the majority of immigrants (around 60%) coming from neighboring countries, especially from Peru and Argentina. This led to a return to the older claims of exceptionalism, namely Chile's non-tropical status.

Probably the most emblematic example of this strategy was the enormous chunk of iceberg that was towed from the Antarctic sea to the 1992 World Fair in Seville, Spain, to serve as the ultimate proof of Chile's lack of *tropicalismo* (cf. Dorfman, 1999). The iceberg was apparently intended to evoke associations of Chile as a cold place that shared not only a climate but also cultural and economic qualities with Northern Europe (Staab & Maher, 2006, p. 105).

The iceberg formed part of a broad campaign of nation-branding to improve the international image of Chile after the dictatorship (Fermandois, 2005, pp. 425-491). This campaign promoted the Chile that the country itself aspired to be: "cool," sober, technically advanced, and efficient—in other words, profoundly modern and, thus, more closely related to England ("the English of the Pacific") or the United States ("the Yankees of South America") than to Bolivia or Brazil (Subercaseaux, 1996). While in the West icebergs are rather associated with wilderness, for Chileans they symbolize a country that "started to jump on the train [of modernity], eager for closer contact with the developed world....The culture of Chilean 'cosmo' began—the aware and travelled Chilean—which was first advanced by the returnees but, over the years, took off on its own" (Contardo, 2008, p. 272).

Advertising, image building, and branding (all modern imaginative technologies) have penetrated all sectors of society, and the state has been one of the first institutions to adapt to this new logic, by becoming the main promoter of Chilean modernity. An important player in this respect is ProChile. This Trade Commission of Chile, responsible for implementing and enhancing Chile's trade policy abroad, was founded in 1974 (the start of the dictatorship) and belongs to the Ministry of Foreign Affairs. Interestingly, one of the first things ProChile did was support the project, *Imagen País* (Country Image). This program, developed by a committee of the government agency for economic development (CORFO) to further develop the image of Chile's role as a member of the globalized world, first focused on the USA, but afterwards broadened its scope to include Europe (specifically Spain and the UK).

Figure 1. Imaginaries of (im)mobility: The old "reflected" in the new...
(Photo: Noel B. Salazar)

Since 2000, the tone and content of ProChile's campaigns have changed considerably. Instead of the triumphalist discourse of modernity and regional exceptionalism, they have increasingly stressed the Latin American character of Chile, focusing on themes such as solidarity and concern for the welfare of the region. This change seems to be the result of pressures from the political Left, which resisted the monolithic approach to Chilean society (exclusively White and Western) of the previous campaigns, as well as of negative reactions from neighboring countries such as Bolivia (Peña, 2003). In 2005, the Imagen País campaign launched the slogan *Chile sorprende, siempre* ("Chile, always surprising"). This catchphrase was used until 2008, when the Fundación Imagen de Chile was created, a new institutional framework designed to provide general coordination of all activities involving Chile's image, both at home and abroad. The foundation's most recent international campaign, launched in September 2010, and entitled *Chile hace bien* ("Chile is good for you"), confirms many of the issues discussed in this chapter.

Contemporary (Im)mobilities

Apart from various migratory movements, rapid developments in the mobility-enhancing technologies of transport and communication have made Chile much more connected to the world than it used to be. In other words, "modernity no longer allows using one's geographical condition as an excuse. Chile is in constant contact with the world" (Stuven, 2007, p. 47). As some argue, "Chile keeps on being an isolated country, but it has been globalizing through communications and, in this sense, the isolation of Chile has become less dramatic, but it keeps on being a relative reality. The isolation expresses itself in that, in general, all references are made towards Chile, inwards" (Duhart, cited in Stuven, 2007, p. 49). Critical observers have noted that others are less homebound than Chileans (Pérez de Arce, 2006, p. 17), or have argued that Chile needs even more outside influence (Gonzalo Rojas, cited in Stuven, 2007, p. 56).

Take the example of Isabel, a young, middle class woman living in a small Chilean coastal town. She frames it as follows: "I have always believed that I won't stay in Chile my entire life. I have thought about living abroad in the near future, but I've never really planned anything.... We Chileans live the life of isolated islanders, don't we?" If Chileans think of their country as an island, it is not only because of Chile's perceived isolation. As Vannini pointed out, another hallmark of islandness is the positive characteristic of insulation, comprising "feelings of protection, safety, distinction, and disconnection" (2011, p. 257). Isabel, for instance, mentions stories she heard from Chileans about being discriminated against in countries like Spain, and comments that she would find it difficult to leave her family behind, being very "attached to the family" ("apegado a la familia"). The latter is something I heard over and over again when talking to Chileans across the country.

From a sedentarist perspective, of course, the absence of geographical mobility is the norm and the ideal. Malkki (1992) showed how the entrenchment of a sedentary worldview naturalizes the link between territory and identity, and in turn, pathologizes territorial displacement. How we interpret this depends on the analytical lens that we use:

> A functionalist perspective would focus on the role of immobility in reproducing the social structure. In a conflict perspective, immobility is a result of the impositions of people or institutions with the power to determine who gets to go and who gets to stay. These divergent understandings might be applied emically—that is, by the people themselves who are defining, imposing, and/or experiencing immobility. (Jónsson, 2011, p. 8)

An analysis that draws on both these perspectives is most enlightening. In the case of Chile, most people I interviewed rely on a conflict view to explain their transnational immobility, while this chapter clearly demonstrates that old imaginaries of (im)mobility play their functional role in keeping people in place and, thus, maintain the status quo in the field of social mobility (see Torche, 2005). Vigh (2009), therefore, proposed "social navigation" as an analytical concept that grants us this double perspective on practice and the intersection between agency, social forces, and change.

It is not all that surprising that Chileans think the way they do, given the long history of images that depict their country as an isolated paradise on earth. Aínsa's description of the Golden Age, where immobility was conceived of as a guarantee of paradise, is enlightening:

> It was simply a matter of being born and dying within the narrow limits of one's own shores, of being satisfied throughout one's life with the products of one's native soil and, above all, of not knowing about or having the curiosity to know about what was outside the precincts of daily life. The felicity of the Golden Age was guaranteed by isolation and self-sufficiency but also by that lack of curiosity toward what might exist beyond the limits of one's own immediate world. The rationalization is simple. If primary needs were satisfied on one's own shore there was no reason to look for new worlds outside the native plot of ground. (1986, p. 28)

The difference nowadays is that Chileans do not need to travel abroad to know what is going on outside their country. If they do not go out and explore the world, the world comes to them, in the form of foreigners (be they migrants or visitors) and global media. According to my informants, Chileans generally look up to foreigners, although they remain selective in which groups they welcome.

As mentioned earlier, Chilean emigration from the end of the 1960s until the beginning of the 1990s was primarily political. In the 1980s, this group of political migrants was expanded by the arrival of postgraduate students (brain drain) and economic migrants from the middle classes (which were quickly becoming poorer). In the 1990s and later, when larger groups of economic migrants left the country, their establishment abroad was facilitated by the communities of Chileans who had preceded them (Yépez del Castillo & Herrera, 2007). Since the 1990s, the percentages of Chileans travelling abroad and the influx of foreigners into the country have dramatically increased. Chileans temporarily leave the country to study, work, or simply travel abroad, being convinced that, in the words of one girl, "travelling opens your eyes and enables you to look at things from a more global perspective." However, "when Chileans travel, they do not particularly strive to know; Chileans arrive abroad and start contacting other Chileans" (Montt, cited in Stuven, 2007, p. 58). In summary,

> the self-referential word is typically Chilean. It does much good for Chileans to leave, but they are afraid of doing so. Not of physically leaving, but of leaving their environment, their frame of reference. The independent type or drifter is very little Chilean. (Duhart, cited in Stuven, 2007, p. 55)

Chileans in exile and returnees play an important role in all of this. They helped many Chileans to get rid of the provincialism they were bound to by trivializing the idea of traveling abroad and by showing that the geographical boundedness can be overcome by technical and economic means. As Contardo stated, "the nouveau riches in Chile, who began to emerge after two decades of sustained economic growth, unveiled a nascent cosmopolitanism that served as a social climbing tool for a small but vocal segment of the population" (2008, p. 272), thus reinforcing the latter argument. Travel became one of the elements relatively easy to access for that segment of the population just below the Chilean upper class. For this class of people the model is not the Chilean *cuico* (the upper class), but the global high society that travels, uses the latest electronic gadgets, eats out in restaurants, and knows about wines. Chileans both inside and outside the country are of the opinion that most career-builders look up to the USA, while Europe sounds "chic" (but is a second option or an ideological first choice). In 2009, for example, 30% of Chileans with a doctoral scholarship went to study in the USA, 18% in Spain, and 17% in the UK (Becas Chile, personal communication).

Holiday-wise, many travel within Chile as it is generally believed that "we have everything in Chile, except tropical beaches" (interview with Solange Fuster, Regional Director of Sernatur). If they have the means to travel outside the country, young Chileans go for the beach or carnival holidays to Brazil, the Caribbean (Dominican Republic), and Mexico (Cancún). Families with children go to Miami, couples travel on round trips to Europe (always including Spain), and the elderly go on international cruises. Those with lesser means go to neighboring Argentina or Peru. Faraway destinations such as China, Canada, and New Zealand are increasingly in demand (Vega, 2011). The newest trend (for those who can afford it) is adventure tourism to the Amazonian rainforest, India, or the Middle East.

Interestingly, one regularly hears arguments against migration. First of all, since Chile is doing relatively well economically within the wider Latin American context, potential migrants are looking beyond the continent's borders. In other words, there is nowadays little incentive to migrate and those wishing to do so need substantial financial resources. Second, although Chile portrays itself as cosmopolitan, in reality the knowledge of foreign languages is very limited (see http://www.trabajando.cl/noticia.cfm?noticiaid=9432). This, in turn, limits migration opportunities. Third, after 9-11 it has become hard to obtain visas and other official migration documents (Cunningham & Heyman, 2004). While these are technical barriers to emigration, there is also a mental barrier: the idea that everything is very far (and consequently expensive, although Chilean people got used to credit). Of course, in the recent past the situation was different. For instance, there were about one million Chileans in (mainly south) Argentina, but many returned back home when Argentina itself was hit by an economic crisis at the turn of the millennium. Nowadays, Chileans travel to Mendoza and Buenos Aires to shop and to show off that Chile is in a better economic condition, and that the Chileans are no longer the *chilenitos* of before.

Conclusion

Culture lies at "the intersection of social experiences and collective imaginaries" (Palet & Velasco, 2002, p. 36). The findings described in this chapter illustrate how imaginative technologies of (im)mobility emerge as sources constitutive of cultural meanings beyond being a mere extension or transfer of them (Salazar, 2010c). Imaginaries are signified and resignified, indicating both socio-cultural continuities and rupture. In the words of Baeza, "social imaginaries are never definitive and, for this reason, the construction of today's reality, for sure, is not the same as the one we will have tomorrow" (2008, p. 288). This applies very well to the representations and practices of (im)mobility. Even when a person is place-bound, his or her imagination can be in movement, travelling to other places and other times. By extension, it can be argued that even when one is in movement, one's imagination can be focused on a singular place (e.g., people in the diaspora re-creating their imagined homeland), and that these imaginaries of fixity can influence one's experience of mobility. How people experience and make sense of transnational (im)mobilities relates to "a greater socio-cultural matrix of values and expectations" (Jónsson, 2011, p. 4). Analyses of mobility are thus best combined with complementary studies of the various "moorings" that promote and constrain it (Hannam et al., 2006). Studying and questioning culture-specific imaginaries of (im)mobility offers us a novel way of grasping the ongoing, global transformations of the human condition (Salazar & Smart, 2011).

The Chilean case nicely shows how constellations from the past can break through into the present in surprising ways. The representational assemblages I discussed above entail particular politics of mobility, "social relations that involve the production and distribution of power" (Cresswell, 2010, p. 21). Jónsson rightly reminded us that "certain forms of immobility may be socially constructed, although they appear as natural, normal and even desirable" (2011, p. 10). The historical continuities of (im)mobilities in Chile are a cultural phenomenon. The constant recycling of old (foreign) imaginaries—a play upon the nostalgic longing for rootedness—serves as a kind of protection. These imaginaries turn out to be a very effective technology of (im)mobility, in the sense that they are used, by citizens and authorities alike, to control and adapt to emerging forms of transnational mobility. As Brubaker argued, "the politics of belonging are generated not by the movement of people across borders, or by the movement of borders across people, but by the *absence* of movement or mobility—in social space, not geographical space" (2010, p. 70). While with the story of the iceberg Chile wanted to show how it could move the unmovable, this chapter illustrates another reality of those who can move but are kept at a standstill by imaginative technologies of (im)mobility.

As Vannini stated, "mobility is *not* inherently desirable" (2011, p. 249). Indeed, the meaning and value attributed to (im)mobility is largely sociocultural. This particular case study illustrates that mobility is not necessarily part of the core of the social imaginary, geopolitics, and cultural life of all of the Americas. The culture of (im)mobility is specific to the region where it is represented and practiced (e.g., Cohen, 2004) and to the specific time period (e.g., Fumerton, 2006). Even if scholars have long since critiqued root-based world views and primordial linkages between people and place (Clifford, 1997; Malkki, 1992), the dominant discourse in Chile keeps on stressing that not being (transnationally) on the move is the quintessential characteristic of what it means to be a true Chilean. Chile keeps thriving on a culture of immobility rather than on one of mobility (at least at the transnational level). The Cordillera and other physical boundaries serve as real barriers for those who stay, and as imaginative horizons for the few who do choose to leave. Immobility persists, as a culturally ingrained and culturally sanctioned response to a rapidly changing world. However, this may only be temporary and not a permanent answer. I concur with Greenblatt et al., who said that "even in places that at first glance are characterized more by homogeneity and stasis than by pluralism and change, cultural circuits facilitating motion are at work" (2009, p. 5). In a global context in which geographical mobility becomes almost normative for any form of achievement, be it economic, academic, or personal, it will become harder and harder to keep Chileans from navigating across boundaries.

Notes

1. All translations in this chapter are my own.

References

Adey, P. (2006). If mobility is everything then it is nothing: Towards a relational politics of (im)mobilities. *Mobilities, 1*(1), 75-94.
Aínsa, F. (1986). From the Golden Age to El Dorado: Metamorphosis of a myth. *Diogenes, 34*(133), 20-46.
Aínsa, F. (1999). *La reconstrucción de la utopía*. Buenos Aires, Argentina: Ediciones Del Sol.
Baeza, M. A. (2008). *Mundo real, mundo imaginario social: Teoría y práctica de sociología profunda*. Santiago, Chile: RIL Editores.
Baritz, L. (1961). The idea of the West. *American Historical Review, 66*(3), 618-640.
Blommaert, J., & Bulcaen, C. (2000). Critical discourse analysis. *Annual Review of Anthropology, 29*(1), 447-466.
Brann, E. T. H. (1991). *The world of the imagination: Sum and substance*. Savage, MD: Rowman & Littlefield.
Brubaker, R. (2010). Migration, membership, and the modern nation-state: Internal and external dimensions of the politics of belonging. *Journal of Interdisciplinary History, 41*(1), 61-78.
Canihuante, G. (2006). *Turismo en Chile: Paisajes y culturas del pasado, presente y futuro*. La Serena, Chile: Fondo Editorial.
Cano, V., & Soffia, M. (2009). Los estudios sobre migración internacional en Chile: Apuntes y comentarios para una agenda de investigación actualizada. *Papeles de Población, 15*(61), 129-167.
Castoriadis, C. (1987). *The imaginary institution of society* (K. Blamey, Trans.). Cambridge, MA: MIT Press.
Clifford, J. (1997). *Routes: Travel and translation in the late twentieth century*. Cambridge, MA: Harvard University Press.
Cohen, J. H. (2004). *The culture of migration in southern Mexico*. Austin, TX: University of Texas Press.
Collier, S., & Sater, W. F. (2004). *A history of Chile, 1808-2002* (2nd ed.). Cambridge, UK: Cambridge University Press.
Contardo, Ó. (2008). *Siútico: Arribismo, abajismo y vida social en Chile*. Santiago, Chile: Vergara.
Cresswell, T. (2010). Towards a politics of mobility. *Environment and Planning D: Society and Space, 28*(1), 17-31.
Cunningham, H., & Heyman, J. (2004). Introduction: Mobilities and enclosures at borders. *Identities: Global Studies in Culture and Power, 11*(3), 289-302.
Doña, C., & Levinson, A. (2004). *Chile: Moving towards a migration policy*. Washington, DC: Migration Policy Institute.
Dorfman, A. (1999). *The nanny and the iceberg*. New York, NY: Farrar, Straus, Giroux.
Fermandois, J. (2005). *Mundo y fin de mundo: Chile en la política mundial 1900-2004*. Santiago, Chile: Ediciones Universidad Católica de Chile.
Franz, C. (2000). Ainogatap, la anti-utopía, o el fin del mundo al revés. *Quehacer, 124*(May-June), 52-58.

Fumerton, P. (2006). *Unsettled: The culture of mobility and the working poor in early modern England.* Chicago, IL: University of Chicago Press.
Greenblatt, S., Županov, I., Meyer-Kalkus, R., Paul, H., Nyíri, P., & Pannewick, F. (Eds.). (2009). *Cultural mobility: A manifesto.* Cambridge, UK: Cambridge University Press.
Hannam, K., Sheller, M., & Urry, J. (2006). Editorial: Mobilities, immobilities, and moorings. *Mobilities, 1*(1), 1-22.
Jedlicki, F. (2001). Les exilés chiliens et l'affaire Pinochet: Retour et transmission de la mémoire. *Cahiers de l'Urmis, 7*(June), 33-51.
Jónsson, G. (2011). Non-migrant, sedentary, immobile, or 'left behind'? Reflections on the absence of migration. *IMI Working Papers* (Vol. 39). Oxford, UK: International Migration Institute.
Lévi-Strauss, C. (1961). *Tristes tropiques* (J. Russell, Trans.). New York, NY: Criterion Books.
Lindquist, J. A. (2009). *The anxieties of mobility: Migration and tourism in the Indonesian borderlands.* Honolulu, HI: University of Hawai'i Press.
Malkki, L. H. (1992). National Geographic: The rooting of peoples and the territorialization of national identity among scholars and refugees. *Cultural Anthropology, 7*(1), 24-44.
Martínez Pizarro, J. (2003). *El encanto de los datos: Sociodemografía de la inmigración en Chile según el censo de 2002.* Santiago, Chile: Naciones Unidas.
Mizón, L. (2001). *Claudio Gay y la formación de la identidad cultural chilena.* Santiago, Chile: Editorial Universitaria.
Palet, A., & Velasco, P. (Eds.). (2002). *Desarrollo humano en Chile 2002.* Vitacura, Chile: Programa de las Naciones Unidas para el Desarrollo.
Peña, J. (2003). Made in Chile: Estilo cultural e imagen-país. *Revista Patrimonio Cultural, 28,* 5-7.
Pérez de Arce, H. (2006). *Los chilenos en su tinto.* Santiago, Chile: Aguilar.
Pizarro, A. (2003). Mitos y construcción del imaginario nacional cotidiano. *Atenea (Concepción), 487,* 103-111.
PNUD. (2003). *Transformaciones culturales e identidad juvenil en Chile* (Vol. 9). Santiago, Chile: Programa de las Naciones Unidas para el Desarrollo.
Roa, A., & Teillier, J. (1994). *La invención de Chile.* Santiago, Chile: Editorial Universitaria.
Sagredo Baeza, R. (2006). *Chile: De 'fines terrae' imperial, a 'copia feliz del edén' republicano.* Paper presented at the Coloquio Internacional Crear la nación. Los nombres de los países de América Latina: Identidades políticas y nacionalismo, Mexico City, Mexico.
Salazar, N. B. (2010a). *Envisioning Eden: Mobilizing imaginaries in tourism and beyond.* Oxford, UK: Berghahn Books.
Salazar, N. B. (2010b). Tanzanian migration imaginaries. *IMI Working Papers* (Vol. 20, pp. 1-29). Oxford, UK: International Migration Institute.
Salazar, N. B. (2010c). Towards an anthropology of cultural mobilities. *Crossings: Journal of Migration and Culture, 1*(1), 53-68.
Salazar, N. B., & Smart, A. (Eds.). (2011). *Anthropological takes on (im)mobility.* Theme issue, Identities: Global Studies in Culture and Power, 18.
Staab, S., & Maher, K. H. (2006). The dual discourse about Peruvian domestic workers in Santiago de Chile: Class, race, and a nationalist project. *Latin American Politics and Society, 48*(1), 87-116.
Strauss, C. (2006). The imaginary. *Anthropological Theory, 6*(3), 322-344.
Stuven, A. M. (2007). *Chile disperso: El país en fragmentos.* Santiago, Chile: Editorial Cuarto Propio.

Subercaseaux, Benjamín. (1973). *Chile, o una loca geografía*. Santiago, Chile: Editorial Universitaria.

Subercaseaux, Bernardo. (1996). *Chile: ¿Un país moderno?* Santiago, Chile: Ediciones B.

Taylor, C. (2004). *Modern social imaginaries*. Durham, NC: Duke University Press.

Torche, F. (2005). Unequal but fluid: Social mobility in Chile in comparative perspective. *American Sociological Review, 70*(3), 422-450.

Vannini, P. (2011). Constellations of ferry (im)mobility: Islandness as the performance and politics of insulation and isolation. *Cultural Geographies, 18*(2), 249-271.

Vega, M. (2011, February 13). Cuánto y dónde viajan los chilenos. *El Mercurio*.

Vigh, H. (2009). Motion squared: A second look at the concept of social navigation. *Anthropological Theory, 9*(4), 419-438.

Yépez del Castillo, I., & Herrera, G. (Eds.). (2007). *Nuevas migraciones latinoamericanas a Europa: Balances y desafíos*. Quito, Ecuador: Facultad Latinoamericana de Ciencias Sociales (FLACSO).

Zavala San Martín, X., & Rojas Venegas, C. (2005). Globalización, procesos migratorios y estado en Chile. In Programa Mujeres y Movimientos Sociales (Ed.), *Migraciones, globalización y género en Argentina y Chile* (pp. 149-191). Buenos Aires, Argentina: Programa Mujeres y Movimientos Sociales.

14

Technology and Technicians Out of Control: The Implementation of Transantiago From a Daily Mobility Point of View

Paola Jirón

On February 10, 2007 Transantiago became the new urban transport system in the city of Santiago, Chile. The new system aimed to modernise public transportation in the city by making the existing, yet complex, network of public buses, or *micros amarillas*, more efficient, comfortable, and safe (Malbran, 2005). The project was financed by international agencies and corporations (including banks and transport consortia). However, the implementation of the modernisation plan proved to be highly problematic and it resulted in travel chaos for hundreds of thousands of urban residents. As a result, Transantiago has been considered to be one of the most disruptive public policy interventions in Chilean history. The Chilean National Congress initiated an investigation to "determine the causes that motivated the failed implementation of a public policy destined in its original idea to substantially improve the transport system, but [which] ended up producing the exact contrary effects" (Comisión Investigadora del Plan Transantiago, 2007, p. 3, author's translation).

Given the complexity of the intervention, Transantiago's longer-term impacts have yet to be adequately assessed. However, by analysing the way in which the plan was conceived and implemented, apparent deficiencies can be identified in the way urban transport experts conceive of urban space as a technical space in which they give limited consideration to the practices of daily urban mobility. By analysing the aims, plans, and decisions that were made prior to its implementation, this chapter explores the problems of decreased public transport coverage, increased waiting time, and the apparent promotion of car use that resulted from Transantiago's implementation. It argues that many of the problems that Transantiago encountered can be attributed to a conceptual deficiency which only provides a partial or incomplete understanding to an everyday life approach to mobility.

The chapter begins by outlining what is meant by an urban, daily mobility approach. It then introduces urban public transport planning in Chile and details the principal transport challenges which prompted the development and implementation of Transantiago. The main aspects of the plan are then presented and the major pre-implementation challenges discussed. The chapter concludes by assessing the ways in which Transantiago affected everyday mobility practices in the city.

An Everyday Life Approach to Urban Mobility

Mobility can be physical, virtual, or imaginative (Sheller & Urry, 2006; Szerszynski & Urry, 2006). Technologies, including televisions, the Internet, and mobile phones facilitate virtual presence in multiple locations simultaneously. However, despite ever-evolving technologies that enable this increasingly distanced, virtual, and/or imaginative mobility, physical co-presence remains a vital component of everyday life for most people and the physical travel that enables this co-presence is "rich and multifaceted" (Urry, 2004, p. 32). Urry (2003) suggested that mobilities are organized into complex patterns that transform the very social relations that the social sciences seek to explain. Thus, being mobile has become a way of life for many and, regardless of any technological advances in communications, physical travel continues to increase.

Although migration, tourism, and residential mobility have specific and extremely complex spatial implications, it is in the mundane daily regimes of urban life that mobility is most frequently expressed. Disentangling and understanding the ways in which urban actors perform mobility on a daily basis has become increasingly important. The study of social interactions in everyday life can help to visualize urban dwellers' lives, including the way power relations are expressed in urban space, as well as the experiences, meanings, and practices of everyday urban living. To talk about everyday life is to talk about the basic sociability of individuals, families, or groups of people as expressed through their daily activities which, themselves, are embedded within existing social structures (Salazar, 1999). The quotidian refers to what people live on a daily basis and is connected to the places where people live, work, consume, relax, relate to others, forge identities, cope with or challenge routine, and establish codes of conduct. It is the daily experiences of being on the bus, of walking, moving, driving, staying put, meeting people, and sharing experiences that form an essential part of "being urban" in contemporary societies.

Individual people perform their daily activities both within and outside their immediate neighborhood and, for many, a large part of their everyday life occurs elsewhere, not only at work, but also in the multiple activities they carry out on a daily basis. As a concept, the everyday life approach emphasizes the interactions that occur between individual practices and social structures, between different kinds of actions, and different levels of consciousness. Everyday practices in urban space serve to mediate interactions between individuals and groups on the one hand, and broader structures and institutions of society on the other (Vaiou & Lykogianni, 2006). A theoretical approach to the everyday can help researchers to examine the invisible and problematic aspects of routine life that are often ignored by mainstream policy makers.

The "everyday" can be understood as events which are otherwise imperceptible and irrevocably lost but which are perhaps the most truly personal (Dewsbury, 2003). These hidden aspects refer to the "secret" parts of people's lives that are either ignored or habitually misinterpreted by urban research and practice as a consequence of being "invisible" or too problematic to include in any analysis (Jarvis, Pratt, & Wu, 2001). Uncovering those aspects, which can remain hidden by abstract quantitative analysis or by qualitative perspectives that enquire on broader understandings of reality, underpins the everyday life approach.

In a reflexive manner, everyday life is constantly changing the lives of urban dwellers in the same way that they change everyday life. The experience of daily living—the quotidian and the daily routines which may appear insignificant to others—is in fact at the core of who we are, what we do, and how we express ourselves. Thus, as space around us changes, so our everyday experience of that space alters. Some changes may occur slowly and we may seldom notice them until we make them part of our daily practices. They may be an accumulation of small changes that, when we suddenly become aware of them and attempt to look back or return, the changes become insurmountable. Others are quick and have instant impact and force us to adapt our daily living accordingly or challenge us to resist them. This everyday living will never be grasped fully, since as soon as it is intercepted it is modified. As Vaiou and Lykogianni (2006) explained, it is in the everyday that changes are recognized and perceived and the potential for change can be found. The need to recognize the importance of grasping this changing logic is only now beginning to be understood.

Lefebvre argued that to "reach reality we must tear away the veil, that veil which is forever being born and reborn of everyday life, and which masks everyday life along with its deepest or loftiest implication" (1991a, p. 57). He insisted on the need to see the activities that others might view as being insignificant. It is in the multifaceted, multitasking moments when everyday life becomes the most vivid or tangible and when people find themselves living more than one life (Ross, 1992, cited in Highmore, 2002). This means that it is precisely when a person is trying to be, for example, simultaneously a mother, a wife, and an employee, that the experience of everyday life becomes most pronounced and profound.

Understanding the daily activities that are performed in urban space is valuable for two reasons. Firstly, the way people experience the city is not often fully considered in urban interventions, as current urban and transport planning practices are mainly informed by abstract information on the city, and provide little recognition as to how the everyday needs to inform policy. Secondly, the everyday could be seen as the closest way to unveil contemporary living and expose the differentiated, multifaceted, and hybrid experiences of mobility and how lives are affected by events, rather than relying on abstract numbers and theories that are often detached from what people do.

Urban Transport in Santiago

According to the 2001 Origin and Destination Survey (SECTRA, 2001), 16 million journeys are made daily in Santiago, of which 42% are performed by car. Prior to Transantiago, the main modes of transport in the city included walking, public buses, bicycles, private automobiles, taxis, *colectivos* (shared, fixed route taxis), the Metro, and suburban rail networks (SECTRA, 2001). Although overall trip numbers increased between 1991 and 2001, the proportion of trips carried out by private cars increased dramatically while bus patronage decreased and the Metro saw only a slight increase in use (SECTRA, 2001). The increased use of private vehicles is closely linked to the decline of bus usage, which generates a vicious circle of car ownership increase, decline in bus patronage, along with a decline in service levels, leading to a lower demand. Increased car use also has negative environmental impacts including greater congestion and atmospheric pollution.

Latin American cities, like most cities around the world, are inequitable spaces. The Chilean capital is no exception. In light of this, any major urban interventions in Santiago need to take this into account as mobility is increasingly seen as an important component in exacerbating urban inequality (Jirón, 2009). According to Zegras and Gakenheimer (2000), just as income is positively correlated with car use, it is negatively correlated with bus use; that is, bus consumption declines as income rises. The car confers a higher degree of social status than the bus, which is "stigmatized as uncomfortable, inconvenient and unsafe in terms of traffic safety and personal security" (p. 33). The modal split, according to Income Groups in Santiago de Chile in 2001 (Figure 1), shows clear differences in transport use. Lower income groups use public buses considerably more than higher income groups, and the latter groups exhibit greater use of the car.

These results are consistent with the 2002 Chilean Census, which indicated that those with the lowest levels of education and those living in greater poverty were considerably less mobile than those with higher income (Delaunay, 2007). Despite the steady rise in car ownership over the past decades, this rise is unevenly distributed. In 2001, the 970,000 cars registered in Santiago were owned by 35% of households (SECTRA, 2001). Low-income households owned 0.2 cars per household, middle income 0.55, and high-income groups owned 1.48 cars per household (SECTRA, 2001; Transantiago, 2004). Trumper (2005) similarly confirmed that although only 4.8% of those households earning over $1.6 million Chilean Pesos[1] monthly do not own a car, 82% of households earning under 280,000[2] Chilean Pesos a month *do not* own a car. Most car owners clearly belong to the highest income groups and lower income households who do own a car tend to operate much older vehicles (on average over 11 years old) that lack catalytic converters, which means that they have greater restrictions placed on travel.[3] Trumper (2005) noted that inequality in Santiago is not only related to inadequate health, education, work, and power for the poor, but also to mobility and inaccessibility. Although uneven daily mobility patterns in Santiago were evident, the aim of Transantiago was to make the system more efficient, and not to address disparities in public transport provision.

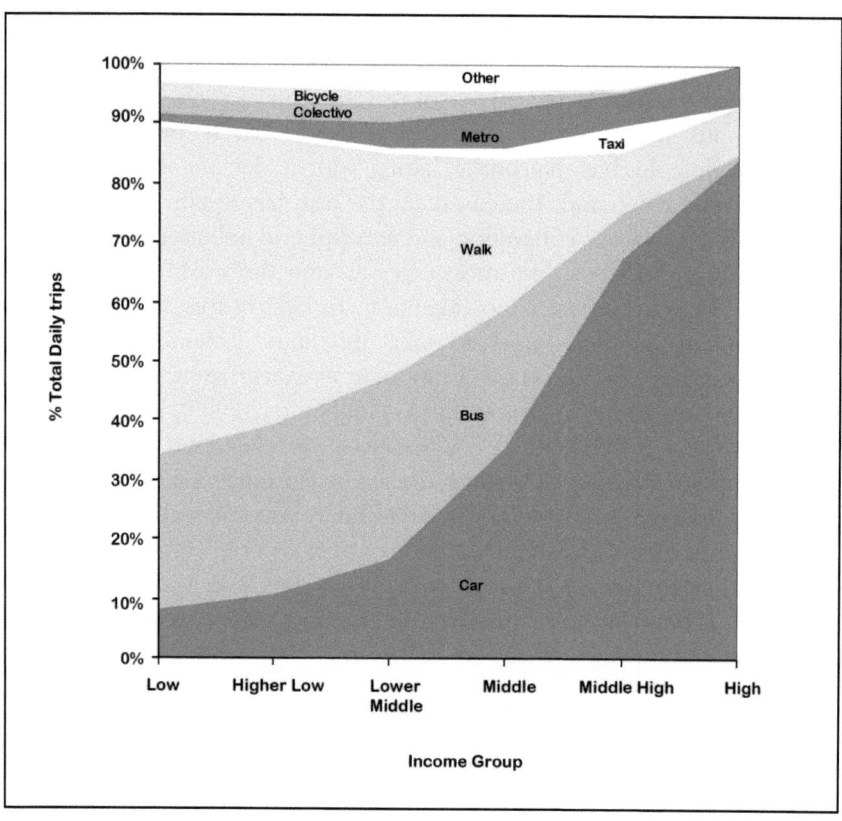

Figure 1. Modal split according to income groups in Santiago de Chile (SECTRA, 2001).

Moreover, bus use decline is closely related to the history of public transport in Santiago, which can be classified in three periods. Up until 1979, the state acted as the regulator of fares, routes, and operators (Transantiago, 2004; Díaz, Gómez-Lobo, & Velasco, 2006). From 1979 until the early-1990s, various initiatives were conducted which eventually resulted in the complete deregulation of urban public transport in the city. Service frequencies and fares were liberalized and restrictions to market entry removed in the expectation that competition would improve the quality of the service and reduce fares. Although liberalization led to an increase in bus supply, it also increased congestion and air pollution and did little to depress fares (Fernández, 2002; Díaz et al., 2006). By the end of the 1980s, high fares, combined with old, uncomfortable, and unreliable buses, resulted in low vehicle occupancy and led to a backlash against liberalization. Increased regulation for road-based public transportation was gradually reintroduced through three packages of measures in 1991, 1994, and 1998. The aim of the concession bids in the mid-1990s was to rationalize, improve, and normalize the public transport system. This induced incremental improvements to the public bus system by reducing the overall number of buses (from approximately 13,000 in 1991, to around 8,000 in 2000), stabilizing fares and improving the quality of the services and routes through a public tendering system (Zegras & Gakenheimer, 2000; Fernández, 2002; Díaz et al., 2006; Orellana, 2006). Although passengers could now reach most areas of the city using a single fare, the newly re-regulated system was far from ideal. It included more than 3,000 independent and private bus owners, each of whom owned an average of two buses and 69% of whom owned fewer than 5 vehicles (Transantiago, 2004; Díaz et al., 2006). This situation led to fragmentation and coordination difficulties.

Prior to Transantiago, the bus system carried approximately 4.7 million passengers a day (SECTRA, 2001) and the average length of a journey per bus was 28 km north-south and 35 km east-west (Díaz et al., 2006). The longest route was 107 km and the shortest 31 km (Transantiago, 2004), meaning that each route crossed the city of Santiago almost completely in one direction or the other. This meant that the capital's main streets were congested while fewer than 50% of seats were occupied during the off-peak period (MOPTT, 2004). Furthermore, over 600 buses crossed the city's main Alameda Avenue per hour (SECTRA, 2001). Most buses went from downtown Santiago to other points in the city. Travellers were able to plan their journey by reaching downtown before connecting to another service, but in most cases passengers often only required one bus to reach their destination. This generated a double role for downtown Santiago as a CBD (Central Business District) and public transport interchange (Krug, 2004).

The sheer volume of buses operating similar routes, combined with the fact that bus drivers' salaries depended on the number of tickets they issued, led bus drivers to compete for passengers. This race for custom generated numerous accidents. Also, the financial precariousness of the individual service providers resulted in poorly maintained and uncomfortable buses that generated high levels of pollution. Finally, there was a serious security issue as drivers who were carrying the daily takings with them on their bus were frequent victims of violent assaults and robbery (Transantiago, 2004).

Although fares were largely affordable to lower-income groups, affordability translated into poor service quality, with the result that higher income groups preferred to travel by car. Bus travellers in Santiago were thus largely at the mercy of bus drivers; buses were dangerous, service was poor and unevenly distributed, and many users experienced difficulty in accessing destinations that were not in downtown areas. Journeys were often difficult, especially for women, those travelling with children, the elderly (who were most vulnerable to theft), and students who were viewed as second class travellers owing to the reduced fares they paid. Overall, satisfaction with service was among the worst of all the evaluated services in Chile (Procalidad, 2002). By the year 2000 most residents, experts, policy makers, and politicians unanimously considered the public transport system to be in urgent need of improvement in terms of service quality. It was clear that unless this situation improved, the bus would only be an option for those with no other transport alternative and the commercial viability of bus enterprises would diminish.

The Transantiago Plan

The idea of Transantiago originated in the year 2000 and was finally implemented in 2007. However, the original plan underwent numerous modifications. At the same time, two other major urban transport interventions were carried out. The Metro was extended and new urban highways constructed. When the new public transport system started taking shape, Metro Santiago was considered to be the backbone of the new public transport system, and significant financial investment and new specialized infrastructure were planned. The Metro development was complemented by the construction of three large, urban highways. These projects were started under the Concessions Program, which aimed to build a network of more than 210 km of toll-paying, urban highways at a cost of US$2 billion (MOPTT, 2004).

Transantiago, meanwhile, involved a complete restructuring of the existing public transport network. It sought to synchronize operations with the existing Metro system through the introduction of an integrated travel fare; install an automatic payment system and a centralized payment collection system; modernize service provision companies' corporate structures; introduce new forms of management and the construction of specialized infrastructure; and promote the "professionalisation" and training of bus drivers (MOPTT, 2004). Ultimately it sought to stabilize and then increase public transport participation in the urban movement through the provision of more coordinated, safer, efficient, and fast mobility (MOPTT, 2004). The system was based on two main concepts: *complementation* and *integration*. The first referred to complementing the use of the bus with the Metro system, while the second involved using a single travel fare for both the buses and the Metro. A new intelligent travel card (Tarjeta Bip!) was introduced on the Metro in 2003 to ease payment and make travelling safer (MOPTT, 2004). It was anticipated that improved integration and coordination between operators and public transport modes would reduce transport expenditure for households living in the periphery and simultaneously improve accessibility (MOPTT, 2004).

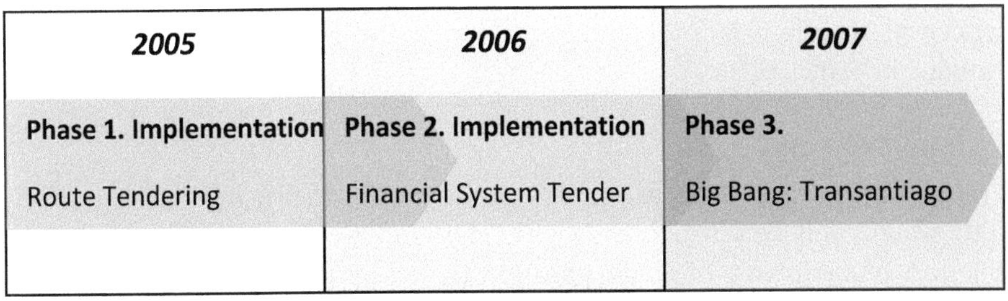

2005	2006	2007
Phase 1. Implementation Route Tendering	Phase 2. Implementation Financial System Tender	Phase 3. Big Bang: Transantiago

Figure 2. Transantiago implementation phases.
Source: Author's elaboration

Transantiago was implemented in three phases. Phase 1, which ran from October 2005 to May 2006, saw public route tendering take place and old buses begin to be replaced by new ones. Phase 2, from May 2006 to August 2006, saw the financial and information systems put out to tender; and Phase 3 represented the "big bang," when the system went into full operation (see Figure 2, Transantiago implementation phases). However, Transantiago started without all the necessary technical, financial, information, and infrastructure systems being in place.

Experts, aware of the magnitude and complexity of the project at hand, expressed concern about how difficult implementation would be and how its success very much depended on community acceptance. Some thought Chileans were quite open to changes and would adapt easily, but many more thought that implementation should have waited until all the systems and infrastructures were in place (Jirón, 2009). Many also agreed that the relation between transport provision and urban planning was very weak and that while the project sought to satisfy existing demands, very little consideration had been given to the future morphology of the city. Transantiago did not include the mechanisms through which the effects of future city expansion on transport infrastructure could be mitigated or evaluated. In addition to this, the institutional situation and the existence of multiple actors made coordinating future planning very difficult. There was also a recognition that decisions were often politically motivated and there was a sense that many opportunities were lost (Jirón, 2009).

The Implementation of Transantiago

In this section, Transantiago is analyzed according to its four main features: transport network, infrastructure, financial systems, and information provision.[4] Each area presented particular problems and although many technical challenges arose upon implementation, the way the plan was conceived reveals a lack of understanding of the complexities of everyday mobility in Santiago.

Transport network

Transantiago's transport network was primarily prepared using data from the 2001 Origin and Destination Survey (EOD)[5] and technical studies (Malbran, 2005). A cost structure of public transport (based on Fernández, de Cea, & de Grange, 2005) and road infrastructure configuration was used to define the routes, frequencies, and capacities of services (Malbran, 2005).

The network was devised as a trunk and feeder system, similar to TransMilenio in Bogotá and RIT in Curitiba. The *trunk* became the Metro and dedicated buses that would extend Metro coverage[6] north, south, east, and west, and *feeders* or local buses would operate within defined zones, feeding passengers into the trunk services. The city was divided into ten feeder zones, each with a designated letter and color (which was displayed on its buses). SECTRA specialists and consultants defined the zones, trunks, feeder routes, and bid specifications (Malbran, 2005). These calculations determined the bus services Transantiago required. Table 1, based on Transantiago's Design Scenario for the year 2005, indicates that the number of services, kilometres, and numbers of buses was going to be considerably reduced.

The system was planned and regulated by the central government, but privately operated through contracts obtained through public tenders for each trunk and feeder zone in 2005. This "authority and multiple operators" organizational model is similar to that used in London and Copenhagen, and it meant that Transantiago would have authority over the design of main routes (trunks) while adjustment to feeder services could be proposed by individual operators (Gschwender, 2005). The bidding system included the provision of modern buses and 15 business units (5 trunks and 10 feeder zones). The feeder zones were designed to collect passengers and transport them to the trunk services quickly, cheaply, and efficiently (MOPTT, 2004). Periphery-to-core transport was prioritized and inter-zone and intra-zone travel was discouraged.

Type of service	Number of services	Km of Services	Fleet (buses equivalent to 80 pax)	
			Peak Morning	Off Peak
Trunk	83	1,853	4,297	1,021
Feeder	114	1,812	986	711
Total	197	3,665	5,283	1,732
Optimised base situation				
'Current' services	370	20,688	8,405	5,085

Figure 3. Transantiago Design Scenario—Year 2005 (based on Malbran, 2005)

The main problems with the network were threefold. Firstly, the heavy reliance on the use of travel models to design the transport network and predict levels of passenger demand ignored the local realities of where, how, and why people move. Secondly, the sudden and complete redefinition of a citywide transport system greatly disrupted people's lives as the change over failed to recognize that cities and mobility practices are social constructs which are built over time. Finally, redesigning the routes of a major public transport system without considering the contours of the urban transport system as a whole demonstrated the tensions and contradictions that are currently present in transport and urban planning systems.

Contemporary transport planning does, of course, involve modeling which is not only valuable but also applicable to cities in Latin America (Gschwender, 2005). In the case of Santiago, the Transantiago development used a model that optimized travel frequencies and considered the total number of potential users and cost of operation.[7] Relying on transport modeling alone presents two main problems for implementing a citywide public transport intervention like Transantiago. Firstly, it presupposes utility maximization and assumes travellers make rational decisions, and secondly, it requires aggregate data of travel patterns to determine the trajectory of optimal routes. However, a closer analysis of how people travel reveals that people do not always seek to minimize travel time or to maximize the value of travel. People often appreciate the time spent on public transport and use it in multiple ways. The purpose of travelling is obviously to get somewhere, but what occurs during the journey as well as prior to and after its completion, is also important.

One of the characteristics of the old bus system was that buses and journey times were slow. Passengers used this time productively by, for example, socializing, relaxing, reflecting, disconnecting, sleeping, studying, working, or dropping off children en route to another destination. By dividing the city into 10 zones that fed into a trunk, the average number of buses individual users had to take to reach their destination increased from only one to two or more. Passengers travelling north to south across the city, for instance, were now obliged to catch one bus from the feeder to the trunk, then take another bus from the trunk to another feeder. This increased the total travel time and fragmented the journey. While Transantiago was designed to make the transport system more efficient and reduce journey times, the need to transfer to connecting services often made the journey more time consuming.

The routes that were designed for the new system also failed to take into consideration the diverse needs of users. For Transantiago's designers, the optimal travel solution was defined in terms of travel time and financial cost, both of which could be reduced under the trunk-and-feeder system. However, the tender specification for each feeder zone did not specify the need to serve each area according to its specific social needs or to optimize the links between adjacent feeder zones. Although designing routes to satisfy every individual travel need is unrealistic and unachievable, the opposite approach is also undesirable. One attempt could have been to ensure access to key local amenities (including hospitals, school, shops, churches, and police stations) within each new feeder zone. However, tender specifications required each service provider to define its optimal route within the zone for which it was bidding. A "perfect route," while difficult to achieve through modeling alone, could have been constructed in conjunction with local residents and transport and urban planning experts. Such an approach may have provided solutions that were more closely tailored to the needs of people living in different areas of the city.

This leads to a second problem with the Transantiago network. Although the previous system was undoubtedly inefficient (and, at times, chaotic), time consuming, uncomfortable, and all too frequently unsafe (due to the number of road traffic accidents and on-board crime), it had developed over a period of time to serve the needs of the city and, as such, it provided near complete, city-wide coverage. The way the new routes were planned, in comparison, failed to recognize that mobility practices are socially constructed over time and space. The new network displaced the existing one and, owing to its complete and sudden modification of the citywide transport system, it interrupted everyday life. Preparation for its launch included only minimal participation from local government, a tier of administration that was arguably best placed to report on local needs and provide information on how routes, destinations, and timetables should be developed.

The disruptive effect the so-called "big bang" implementation of Transantiago had was not too dissimilar to that of the launch of a new national currency. However, given the complexity of the system and the general lack of citizen preparation for its launch, a phased, gradual implementation would arguably have been preferable. Although a trial period was scheduled, it merely involved introducing new buses onto existing routes. Experiences of other transport schemes in other Latin American cities, like TransMilenio in Bogotá[8] or RIT in Curitiba, provided positive examples of phased implementation. Such an approach would have allowed for a more rapid acceptance and seamless adoption of the system.

A third problem involved conceiving Transantiago solely in terms of public transport. Indeed, other transport modes, including cars, bicycles, *colectivos*, taxis, and walking, were not included within its remit. Transantiago attempted to reduce the existing bus fleet from 8,000 to 4,500 vehicles, where trunk routes would travel 36 km and local ones 18 km (Malbran, 2005). This would eliminate previous inefficiencies of a large percentage of buses travelling at less than optimal capacity over the same main streets. However, no consideration was given to the effect that reducing bus capacity and service frequency by at least 25% would have on other modes.

Clearly, urban transport inefficiencies are not solely due to public transport; private vehicles also play a major role. Understanding urban mobility involves seeing all movement in the city: people and goods, over and underground, public and private, motorized and non-motorized, and inter-urban and rural, that run to, from, and through the city. Transantiago failed to consider the varied experiences people have of using different modes of transport—particularly car driving, walking, cycling, and riding in *colectivos*— and neglected to see public transport as complementary to these. Assuming that people will automatically get out of their cars and switch to using a new public transport system appears naïve in the extreme. Individual travel choice is based on decisions regarding convenience, cost, comfort, and a myriad of other subjective attributes, and is dependent not only on one's ability to use or access a particular mode, but also on the relative merits of one mode vis-à-vis another. Car-sharing or car-pooling takes place informally and can be an efficient way of moving around the capital, but as a practice it is not officially recognized, discussed, or encouraged (ASINTRA, 1997).

The initiative to implement Transantiago and simultaneously build an urban highway system appears contradictory. Indeed, while public transport was being encouraged, opportunities for private transport were simultaneously being enhanced. For some experts, a city the size and population of Santiago requires urban highways and these are indicative of the level of development of the country (Echeñique, 2006). If so, and if implementation of both systems was inevitable, then effort had to be made to reconcile their objectives and make Transantiago as effective an alternative for car users as possible. This means making routes adequate for car drivers, understanding secondary issues of car use, including parking, but also restricting its use where public transport could offer a viable alternative.

Infrastructure

Infrastructure for Transantiago included road maintenance and improvement, a 40 km Metro extension, construction of segregated corridors for buses and cars, new road connections to decongest specific arteries, intermodal exchange stations, new bus stops, and 35 transfer stations. These infrastructure developments were carried out in 12 stages by the Servicio de Vivienda y Urbanismo (Housing and Urbanism Service) (SERVIU, 2007), and MOP (2007) through the Concessions Programme. The first stage commenced in 2004 and was completed largely on time. The second stage was to start in 2007 and include other main road corridors and maintenance of various routes. Subsequent phases were to be launched between 2007 and 2010 and would include over 41 km of road improvement (López, 2007).

It was clear that the plan required a great deal of new transport infrastructure and significant financial investment. The Metro was not being used to its full capacity because it did not reach many areas of the city and it cost more than the buses. To overcome this problem, Transantiago inaugurated a new, single fare regime which included both buses and the Metro. It was predicted that this change would increase use of the Metro by 250% (Tomic, 2006). However, few initiatives were taken to minimise the impact of such an increase. Shortly after implementation, overcrowding on the Metro became a serious safety concern.

Other infrastructure problems were also related to their less than optimal design, including road width, bus stops, and access to these bus stops. The new transport system was not conceived as an opportunity to improve Santiago's infrastructure, rather it was a carried out as a way of smoothing out bus flows. Transantiago could have been an opportunity to improve access to transport areas by building more attractive walkways, introducing green areas, lighting, and road protection bars, as well as bus stops that would protect waiting passengers from the rain and make buses more accessible to those with reduced mobility. Such areas could have been linked to other transport facilities, such as drop off places for cars, *colectivo* stands, bicycle parking, cycle routes, and segregated *colectivo*-only routes. A vision of how both transport and urban areas complement each other would have had an important contribution to the city as a whole; but restricted funds, limited time, and the particular remits of the organizations involved in Transantiago's delivery meant that such opportunities went unfulfilled.

Implementing such a radical change in daily living required all necessary infrastructure to be ready on time, yet it was not. Delays in completing the necessary infrastructure were not only due to implementation difficulties, but also to how the infrastructure investment was prioritized. Transantiago required certain standards and capacity to be met and so considerable overground investment had to be undertaken to improve existing roads and public spaces. However, the Metro (rather than Transantiago) received the priority investment (Vallejos, 2006).

Similar schemes in Curitiba and Bogotá demonstrated that improvements to public transport provision could be carried out without building a new Metro system, and at much lower cost, through well-designed, integrated bus systems (Wittig, 2006). The decision to build new Metro lines was logical if the Metro is considered on its own, but not when it is viewed as part of a system. Yet, because such urban investments are complementary to each other and have multiple impacts, as soon as developments to other transport systems are undertaken, the benefits and costs associated with other schemes may no longer be justified. As a tool, urban investment evaluation requires urgent revision in the future.

In Chile, there have been major discussions with reference to the contradiction between the provision of urban highways and public transport (Cruz, 2002; Martínez, 2002; Delpiano, 2006; Echeñique, 2006; Giesen, 2006). The prevailing discourse is that people use cars and will continue to do so, so until public transport becomes a viable or feasible option, car circulation needs to be improved. These investment decisions rarely include other costs associated with having more cars on the road, including parking, congestion, and pollution. However, urban highways decisions, plans, and investments in Santiago were decided separately and without fully considering the impact one would have on the other.

Financial System

One of the most innovative solutions to improve transport efficiency and safety in Santiago was the introduction of an integrated payment system for public buses and the Metro. Integrating both fare systems was necessary given the inevitable increase in modal transfers. Fare integration involved designing a new payment collection system, new fare controls, and new methods of financial administration and income distribution. The Transantiago Financial Administrator (Administrador Financiero Transantiago, AFT) was created with the aim of administering this new financial system.

The international bidding system involved a complex definition of fare setting that stipulated that fares must not be overly increased. Fares would be collected through an "intelligent" card, which would be activated upon boarding a bus or entering the Metro. The AFT would administer revenues and payment to transport companies depending on the number of buses on the road and frequency, through a combined system of Fare + Salaries + Transport Service + Support System Salary + Infrastructure Payment. Transport profit for operators would be made according to the transport service provided in terms of the number of passengers transported.[9] AFT was to issue a sole access means, produce and distribute the intelligent touch cards, provide a top-up network, administer funds and credit provision, and operate the bus control system (MOPTT, 2004). The main problems with this financial regime were twofold: an operational one related to not recognizing the importance of travellers quickly adapting to a radically new system; and a more profound and perhaps ideological one relating to a private system operating without any form of public subsidy.

The first problem relates to the public acceptance of the new system and the new technology. As previously discussed, the majority of public transport users in Santiago are lower-income groups, who may have difficulty in producing sufficient immediate cash for everyday living and normally dispose of travelling money on a daily basis. The new card system required topping up, and travellers would have to organize themselves to charge their cards on a weekly or daily basis. At the beginning, card charging was only available in designated Metro stations and some large shopping areas and these locations were not easily accessible to all public transport users. This oversight inevitably led to implementation problems and the designated trial month was too short to highlight and solve all the problems with the system prior to its "big bang" launch.

Fare integration was one of the best solutions offered by Transantiago, as with a single fare, travellers could use different bus routes as well as the Metro system. However, those travelling the longest distances with most transfers would end up paying more than those travelling shorter distances and transferring less, as a traveller within a single zone would pay less than one who made three transfers to cross the city (and who was more likely to be a lower-income traveller living on the city's periphery). Little analysis of this was done, mainly because transport efficiency and not social equity was the main aim of the system. However, although the prices would eventually be increased, the system was conceived without any subsidies and all the funding to operate Transantiago was expected to come from user fares. This disposition was not restricted to transport but was a national economic one, based on the neo-liberal logic of Chilean public policies which dictate that subsidies are to be avoided in service provision where users can pay. Since the aims of the Plan were not to improve social inclusion or reduce urban inequality, subsidies were viewed as being incompatible with the system.

Some of the consequences of existing mobility practices include uneven access to people, activities, and places; unequal opportunities in the city; and the creation of tunnel-like flows where people do not meet, leading time-poor yet cash-rich groups to opt for easier or alternative solutions which would distance them further from lower-income groups, thereby further fragmenting the city. Transantiago presented Santiago's urban dwellers the potential to improve urban accessibility and make it more equitable. One way could have been by subsidizing the demand for those experiencing most difficult conditions or exclusion. However, under the no-subsidies policy, price increases will be almost inevitable, further exacerbating inaccessibility and perhaps widening social inequity.

Information System

The Transantiago Consumer Attention Service (Servicio de Atención a Usuarios de Transantiago, SIAUT) was designed to provide route information to transport service operators and travellers. Aside from collecting, storing, and distributing operational information and producing reports to fulfill service contracts, it also had to provide users with information. It was in charge of promoting, training, and educating citizens about the new integrated transport system for the city (Transantiago, 2007). It was to have a customer service office and a telephone hotline which would provide users with information on timetables, routes, and live service information. This would be achieved through coordination with the Transit Control Operating Unit (Unidad Operativa de Control de Tránsito, UOCT) who monitor traffic flow and congestion in the city. In order to introduce users to the new system, a Guide of Transantiago, containing route maps and information, was published one month prior to its implementation. However, as SIAUT was not fully operational at this time, operating difficulties were experienced from the outset.

The information system was put out to public tender approximately nine months prior to implementation, quite late, given the task at hand. The nine months were mainly used to set up the system; the actual schemes only started operating approximately one month prior to full implementation. The communication strategy used was similar to a marketing one and aimed to sell the Transantiago "product" rather than educate the population or provide an effective communication strategy. This is not surprising given the emphasis on economic and technical criteria to evaluate the Plan and lack of consideration of the "social" and daily life implications in the way the Plan was conceived.

This emphasis overlooked the fact that significant and definite changes were going to occur: bus frequencies were going to decrease, fewer bus stops would be served, and payment format was going to change. Previous requests of "¿me lleva por cien?" ("Will you take me for $100?"(approximately 10 cents)) to the driver as a way of paying less to travel shorter distances, would be eliminated. The imminent cultural changes called for more time and preparation. These changes required informed discussions, with at least one year of anticipation,[10] using educational messages to prepare future users for it.

The cultural change required designing a communication strategy that would not only inform, but also receive and distribute accurate information. Public campaigns alerting the population about what was coming were limited and delayed to one month prior to implementation, and the information provided was not clear. There was insufficient information to warn and prepare passengers of what was coming, not only in terms of preparing for the radical change in routes, but also in terms of how the integrated fare system was going to operate, and the cost of each journey.

Given that this represented such a radical cultural modification of daily practices, it required understanding the city as a social and cultural construct and incorporating this construct into the Plan from its inception. The preparation of Transantiago as a social construct would have also involved a participatory process to understand and incorporate the dynamism of everyday practices. This implied understanding local idiosyncrasy and incorporating a dynamic participation and consultation from the outset of the Plan. Dynamic public discussions regarding the system were never held, although proposals were made to have community organizations represented in the discussion (Jirón, 2009). Professional and academic presentations were held, but very few were open to public debates. This could have been translated into simple, clear, and multiple forms of transport education, using, for instance, local personalities that people would identify with, to demonstrate the use of the new system, the card, new ways of boarding, and new routes. This would have helped to establish a dialogue with urban travellers about what was coming, in order to make it part of the local language and social construct. There are various techniques, methodologies, and experiences where this has been successfully incorporated, but it required the will to do it.

As it was, people were uninformed and the vague information that was provided generated fear and resentment, causing the system to begin losing legitimacy before it was even implemented. Formulating a design as complex as Transantiago, and with such a significant impact, required at least the partial involvement of those who actually needed to use the public transport system, as opposed to experts designing it from computers, far away from the streets.

Conclusion

Transantiago's first year of operation was described as "chaotic," a "nightmare" (Long, 2007; El Mercurio, 2008; The Economist, 2008), and "one of the worst disasters of public policies that could have happened to a government"[11] (Pérez Yoma, cited in Alamo, 2008). The main manifestations of the problems included an insufficient number of buses on the streets, inadequate route definition, unfinished infrastructure, breach of contracts, as well as payment and control systems failure. Collectively, these caused people to experience extreme difficulties in adapting to a new system. This situation exacerbated travel and waiting time, overcrowding, and complaints, and generated considerable difficulties for citizens who relied on public transport to move around the city. All of this had significant consequences in the daily activities people undertake. The turmoil with such a major change could perhaps be understandable for the first few months, but four years on, many of the problems are still being experienced and arguably have yet to be adequately addressed.

Some of the modifications have been positive, for instance having a single payment card, yet payment evasion remains a problem. Infrastructure is still being built but problems have been detected in terms of infrastructure and urban design as well as legibility. Routes are still being modified. People tend to accommodate, but many of the old practices that were meant to be eradicated are still present in today's Transantiago: fare evasion, congestion, traffic accidents, and pollution.

This chapter has argued that the main problems associated with Transantiago's implementation were foreseeable and originated from a limited understanding of urban daily mobility from the user's point of view. As Hine and Mitchell (2001, p. 330) explained, "the notion of the universal, disembodied subject which has shaped transport policy fails to present individuals as participants in a range of activities across different locations." It has also created top-down interventions that, apart from making life difficult for urban residents, simply do not have the effect that the models anticipated, as people do not always behave the way these models predict.

The experience of Transantiago illustrates the major challenges associated with approaching urban transport interventions from the basis of abstract models and what happens when planners have little conception of how transport systems are used, experienced, and embedded in everyday lives. The problems detected in the Transantiago implementation were threefold:

1. An inadequate understanding of how individuals experience mobility in the city.

2. A failure to acknowledge the complexity of social relations underlying this mobility.

3. A top-down planning intervention which failed to involve meaningful public consultation or participation.

The results were fragmented urban management practices, serious social and environmental impacts, and disruption to the everyday lives of city residents.

The limitations of technicians and technology to "solve" urban transport problems without consulting the users of these systems need to be recognized. This involves recognizing that urban mobility problems are highly complex and thus require better understanding of their everyday implications. Furthermore, limitations in the way this plan was conceived demonstrate the need for research that is better attuned to the diversity of urban experiences, the way social constructs develop, and the ways in which the city is linked not only by transport but also by the relations people have with space, and the ways in which space impacts on these relations.

Notes

1. Approximately £1,600 monthly (CASEN, 2003).

2. Approximately £280 monthly (CASEN, 2003).

3. Because of high levels of air pollution, Santiago restricts the use of private cars between March and December. This restriction covers vehicles that are not fitted with a catalytic converter. On pre-emergency and emergency days, when air pollution is particularly high, all cars are restricted.

4. The institutional framework also presented major problems in the implementation of Transantiago.
5. Main information gathered by EOD refers to data from households, individuals, and trips regarding: origin and destination, trip temporal distribution, modal partition, trip purpose, trip duration, freight origin and destination, flow on main streets, circulation speed, and socioeconomic characteristics of travellers (SECTRA, 2004).
6. The cost of this was US$2, billion while the over-ground infrastructure was US$200 million.
7. For details see Malbran (2005) and Fernández et al. (2005).
8. Transantiago was partially inspired by TransMilenio, the metropolitan transport system implemented in Bogotá from the year 2000. Implementation for TransMilenio was done in stages, with the new system operating in parallel to the existing one for a time (López, 2007).
9. Although originally proposed in terms of kilometres travelled, as in Bogotá.
10. TransMilenio, a more gradual initiative, started its communication strategy one year prior to implementation.
11. Author's translation. Edmundo Pérez Yoma is currently the Minister of the Interior.

References

Alamo, C. (2008). Edmundo Pérez Yoma: 'Hay que estar preparado para perder el poder.' *Revista Cosas, 32,* 42-45.

ASINTRA Ltda. (1997). *Estudio: Mediciones de tránsito evaluación auto compartido comuna de Providencia,* ASINTRA Ltda.

CASEN. (2003). *Encuesta de Caracterización Socioeconómica Nacional.* Santiago, Chile: Ministerio de Desarrollo Social.

Comisión Investigadora del Plan Transantiago. (2007). *Propuesta conclusiones.* Comisión Investigadora Encargada de Analizar los Errores Cometidos en el Proceso de Diseño e Implementación del Plan Transantiago. Santiago, Chile: Cámara de Diputados de Chile.

Cruz, C. (2002). Sin problemas de convivencia. Auto versus transporte público II. *Revista Universitaria, 78,* 67-72.

Delaunay, D. (2007). Relaciones entre pobreza, migración y movilidad: Dimensiones territorial y contextual. *Revista Notas de Población, 84,* 87-130.

Delpiano, M. O. (2006). Marcial Echeñique: Movilidad es riqueza. *Revista Foco 76, 2,* 11-15.

Dewsbury, J.-D. (2003). Witnessing space: 'Knowledge without contemplation.' *Environment and Planning A, 35*(11), 1907-1932.

Díaz, G., Gómez-Lobo, A., & Velasco, A. (2006). *Micros en Santiago: De enemigo público a servicio público.* In A. Galetovic, I. Poduje, & A. Aravena, (Eds.), *Santiago: Dónde estamos y hacia dónde vamos* (pp. 425-460). Santiago, Chile: Centro de Estudios Públicos.

Echeñique, M. (2006). *Las vías expresas urbanas: Que tal rentables son?* In A. Galetovic, I. Poduje, & A. Aravena (Eds.), *Santiago: Dónde estamos y hacia dónde vamos* (pp. 461-488). Santiago, Chile: Centro de Estudios Públicos.

The Economist. (2008, February 7). The slow lane: Fallout from a botched transport reform. Retrieved from http://www.economist.com/world/la/displaystory.cfm?story_id=10650631

El Mercurio (2008, February). 1er aniversario de Transantiago. *El Mercurio.*

Fernández, J. E. (2002). Hacia una ciudad sin tacos. Plan de transporte urbano para Santiago 2010 (II). *Revista Universitaria, 78,* 49-51.

Fernández, J. E., de Cea, J., & de Grange, L. (2005). Production costs, congestion, scope and scale economies in urban bus transportation corridors. *Transportation Research Part A, 39*(5): 383-403.

Giesen, E. (2006, August). *Transantiago: Un análisis crítico.* Paper presented at the Taller sobre Ciudad Sustentable Conference, CEPAL, Universidad de Chile, Santiago, Chile.

Gschwender, A. (2005). *Improving the urban public transport in developing countries: The design of a new integrated system in Santiago de Chile.* Paper presented at the Ninth Conference of Competition and Ownership in Land Transport, Lisbon, Portugal.

Highmore, B. (2002). *The everyday life reader.* London, UK: Routledge.

Hine, J., & Mitchell, F. (2001). Better for everyone? Travel experiences and transport exclusion. *Urban Studies, 38*(2), 319-332.

Jarvis, H., Pratt, A. C., & Wu, P. (2001). *The secret life of cities: The social reproduction of everyday life.* New York, NY: Pearson.

Jirón, P. (2009) *Mobility on the move: Examining urban daily mobility practices in Santiago de Chile.* (Unpublished doctoral dissertation). London School of Economics and Political Science, London, UK.

Krug, C. (2004). El transporte urbano y sus alternativas para Santiago de Chile. *Urbano, 6*(7), 20-23.

Lefebvre, H. (1991). *Critique of everyday life* (Vol. 1). London, UK: Verso.

Long, G. (2007, December 14). The mass transit system from hell. *Time.*

López, C. (2007). Problemas de infrastructura: Las tres deudas del Transantiago. *En Concreto, 55,*12-17.

Malbran, H. (2005). Transantiago: Un nuevo sistema de transporte público para Santiago de Chile. Paper presented at the *1er Congreso Internacional de Transporte Sustentable, Sistema Integrado de Transporte—BRT* Conference, Mexico City, Mexico.

Martínez, F. (2002). Evitar un Santiago 'automovilizado': Autos versus transporte público (III). *Revista Universitaria,78,* 70-72.

MOP (2007). *Síntesis regional 2007: Región metropolitana.* Santiago, Chile: Dirección de Planeamiento, Ministerio de Obras Publicas.

MOPTT (2004). *Plan de transporte urbano de Santiago, Chile: Transantiago subete.* Santiago, Chile: Ministerio de Obras Publicas, Transporte y Telecomunicaciones, Gobierno de Chile.

Orellana, A. (2006). *Una gobernanza metropolitana para el Transantiago: Dilema no resuelto.* Paper presented at the Ciudades Sustentables: Propuestas para la Gestión Estratégica Conference, CEPAL, Santiago, Chile.

Procalidad (2002). *Indice nacional de satisfaccion de consumidores, 1er semester.* Santiago, Chile: Procalidad.

Salazar, C. E. (1999). *Espacio y vida cotidiana en la ciudad de México.* México City, D. F., Mexico: Centro de Estudios Demográficos y de Desarrollo Urbano, El Colegio de México.

SECTRA (Secretaría de Planificacíon de Transporte). (2001). *Encuesta origen destino de viajes.* Retrieved from http://sintia.sectra.cl/ (no longer accessible).

SECTRA (Secretaría de Planificacíon de Transporte). (2004). *Requerimientos de información en la planificación de transporte.* Santiago, Chile: Ministerio de Transportes y Telecomunicaciones.

SERVIU (Servicio de Vivienda y Urbanización Metropolitano). (2007). Subdirección de pavimentación y obras viales. Retrieved from http://pavimentacion.serviurm.cl/

Sheller, M., & Urry, J. (2006). The new mobilities paradigm. *Environment and Planning A, 38*(2), 207-226.

Szerszynski, B., & Urry, J., (2006). Visuality, mobility and the cosmopolitan: Inhabiting the world from afar. *British Journal of Sociology, 57*(1): 113-131.

Tomic, B. (2006). D No 51. S. C. d. M. Transantiago. (Confidential internal letter). Santiago, Chile: Metro de Santiago.

Transantiago. (2004). *Diagnostico del sistema de transporte público.* Santiago, Chile: Ministerio de Obras Publicas Transporte y Telecomunicaciones.

Transantiago. (2007, November 27). *Transantiago informa.* Santiago, Chile: Ministerio de Obras Publicas Transporte y Telecomunicaciones.

Trumper, R. (2005). Automoviles y microbuses: Construyendo neoliberalismo en Santiago de Chile. In R. Hidalgo, R. Trumper, & A. Borsdorf (Eds.), *Transformaciones urbanas y procesos territoriales: Lecturas del nuevo dibujo de la ciudad latinoamericana* (pp. 71-82). Santiago, Chile: GEOLibros, Pontificia Universidad Católica.

Urry, J. (2003). Social networks, travel and talk. *British Journal of Sociology, 54*(2), 155-175.

Urry, J. (2004). Connections. *Environment and Planning D: Society and Space, 22*(1), 27-37.

Vaiou, D., & Lykogianni, R. (2006). Women, neighbourhoods and everyday life. *Urban Studies, 43*(4), 731-743.

Vallejos, D. (2006). Entrevista a Germán Correa. Transporte en Santiago: Un proyecto controvertido. *Revista de Arquitectura 13*, 32-40.

Wittig, A. (2006). Germán Correa y el Transantiago: La lacra de no ser negocio. *Revista Foco 76, 2*, 28-33.

Zegras, C., & Gakenheimer, R. (2000). *Urban growth management for mobility: The case of the Santiago, Chile metropolitan region.* Prepared for the Lincoln Institute of Land Policy and the Cooperative Mobility Program. Cambridge, MA: MIT Press.

About the Contributors

Lucy Budd is a Lecturer in Transport Studies in the Department of Civil and Building Engineering at Loughborough University in the United Kingdom, who specializes in commercial aviation. Her main research interests include the geographies of airspace, the historical development of civil aviation, and the socio-cultural dimensions of aeromobility in all their various manifestations. She has worked on a number of major research projects and has published in a wide range of academic and industry journals. Recent academic publications include articles and chapters in *Progress in Human Geography*, *Aeromobilities*, *The cultures of alternative mobilities: Routes less travelled*, *Environment and Planning*, *Political Geography*, *International Business Travel in the Global Economy*, *Journal of Transport Geography*, and *Health and Place*.

Jim Conley is an Associate Professor in the Department of Sociology at Trent University, Peterborough, Ontario, where has been teaching a course on the sociology of the automobile since 2003. He is the co-editor (with Arlene Tigar McLaren) of *Car Troubles: Critical Studies of Automobility and Auto-Mobility* (Farnham, UK: Ashgate, 2009).

Barbara Crow is the Associate Dean of Research, Faculty of Liberal Arts & Professional Studies, York University, Ontario, Canada. Dr. Crow's current research interests relate to the social, cultural, political and economic implications of digital technologies. Her most recent project, which she is undertaking in collaboration with Professor Kim Sawchuk of Concordia University and is funded with a SSHRC Standard Research Grant, focuses on senior citizens and mobile technologies. She has also edited collections on mobile technologies, including: *The Wireless Spectrum: The Politics, Practices and Poetics of Mobile Communication* (UTP, 2010), co-edited with Michael Longford and Kim Sawchuk; a special issue in 2008 of the *Canadian Journal of Communication* entitled "Wireless Technologies, Mobile Practices," co-edited with Kim Sawchuk and Richard Smith; and a special issue in 2008 of *Atlantis* entitled "Digital Feminisms," co-edited with Sheila Petty. She has also edited several book collections, including *Radical Feminism: A Documentary Reader* (NYUP, 2000) and three editions of *Open Boundaries: A Canadian Women's Studies Reader* (2000, 2004 and 2008) with Lise Gotell.

Rhys Evans is Associate Professor of Rural and Community Development at the *Høgskulen for landbruk og bygdenæringar* – the Norwegian University College of Agriculture and Rural Industries, in Rogalande, Norway. He is the co-founder of the Equine Research Network, an international network of social science researchers with interests in human-horse relations and all matters equine. He spent a decade in Scotland working in the field of rural community development with an emphasis on Community Woodlands and Asset-based rural community development. Prior to entering academia, he was employed as a truck driver in Canada and the USA, where he experienced every type of trucking from the Ice Roads in northern Canada, to gypsy trucking across the American South. He also operates a private consultancy, Integrate Consulting, which specializes in multi-method social research, policy analysis, evaluation and training. He is a human geographer by training, and by inclination --as he has always been interested in going to new places and 'sussing them out'. His publications have appeared in such sources as *Animals in Place: Critical animal studies*, *Geographies of Rhythm: Nature, Place, Mobilities and Bodies*, *Researching Sustainability: Social science methods, practice and engagement,* and *Spaces of Masculinity*.

Christian Fisker is a Doctoral Student in the Department of Architecture, Design and Media Technology at Aalborg University in Denmark. Christian's recent research has been exploring seniors' (auto)mobility and how various life connections are configured and reconfigured when the ability to drive a car is diminished or lost completely. Christian has been a community planner, a writer, a part-time instructor at Ryerson University in environmental gerontology and computer and Internet courses for seniors, a demographic and market research consultant, a seniors' housing developer, a senior policy advisor and chief of staff to an Ontario cabinet minister in seniors' issues, long-term care and tourism portfolios.

Jennie Germann Molz is Assistant Professor of Sociology at the College of the Holy Cross in Worcester, Massachusetts, USA. Her current research interests revolve around tourism mobilities, mobile technologies and social media. She co-edited, with Sarah Gibson, *Mobilizing Hospitality: The Ethics of Social Relations in a Mobile World* (Ashgate, 2007) and has published articles in journals such as *Body & Society, Citizenship Studies, Environment and Planning A, Journal of Tourism and Cultural Change, Space and Culture,* and *Tourism Geographies*. She has also contributed chapters to several edited collections including *Culinary Tourism, Tourism Mobilities, Emotional Geographies, Tourism and Social Identities,* and *Mobile Methods*.

Ole B. Jensen is Professor of Urban Theory in the Department of Architecture, Design and Media Technology, at Aalborg University, Denmark. He is co-founder and board member of the "Centre for Mobility and Urban

Studies" at Aalborg University. Key mobilities publications articles and chapters in *The Contemporary Goffman*, *Mobilities*, *Swiss Journal of Sociology*, and *Making European Space: Mobility, Power and Territorial Identity*.

Paola Jirón is a Chilean academic from the Institute of Housing, Faculty of Architecture and Urbanism at the University of Chile, where she coordinates the Masters Programme on Residential Habitat. She has carried out extensive research, teaching and consultancy work in the areas of housing, urban quality of life, gender and urban mobility practices. At the moment she is undertaking a FONDECYT funded research on social exclusion and urban daily mobility in Santiago. She has been a member of the Advisory Board of Human Settlement Network organised by United Nations Habitat Programme. Her publications have appeared in such sources as *Environment and Urbanisation*, *Swiss Journal of Sociology*, and *International Political Sociology*.

Noel B. Salazar obtained his Ph.D. from the Department of Anthropology at the University of Pennsylvania and is currently FWO Research Fellow at the Faculty of Social Sciences, University of Leuven. His research interests include anthropologies of mobility, the local-to-global nexus, discourses and imaginaries of Otherness, cultural brokering and cosmopolitanism. Recent publications include articles and chapters in *Identities: Global Studies in Culture and Power*, *Crossings: Journal of Migration and Culture*, *Heritage and Globalization*, *International Journal of Tourism Anthropology*, and *Annals of Tourism Research*. He is the author of *Envisioning Eden: Mobilizing Imaginaries in Tourism and Beyond* (Berghahn Books, 2010).

Kim Sawchuk is a Professor in the Department of Communication Studies, Concordia University, Quebec, Canada and the Editor of the *Canadian Journal of Communication*. Dr. Sawchuk current research interests address how various, and often unrecognized, communities of practice negotiate agency and power in the digital world. Her long-standing research interests concern the politics of embodiment from a feminist perspective. Her most recent project, which she is undertaking in collaboration with Dr. Barbara Crow of York University and is funded with a SSHRC Standard Research Grant, focuses on senior citizens and cellphone technology. She has also edited collections on mobile technologies, including *The Wireless Spectrum: The Politics, Practices and Poetics of Mobile Communication* (UTP, 2010), co-edited with Michael Longford and Barbara Crow; a special issue in 2008 of the *Canadian Journal of Communication* entitled "Wireless Technologies, Mobile Practices," co-edited with Barbara Crow and Richard Smith. She has also co-edited several book collections, including *When Pain Strikes* (1999), *Wild Science: Reading Feminism, Medicine and the Media* (2000), *Embodiment* (2007), and *USED/Goods* (2009).

Nick Scott is a Ph.D. candidate in the Sociology and Anthropology Department at Carleton University, Ontario, Canada. His research focuses on everyday travel, especially automobility and velomobility, and its relationship with science, technology, and citizenship. His research has appeared in such publications as *ARC: The Journal of the Faculty of Religious Studies McGill University* and in edited books such as *Mobilizations, Protests and Engagements: Canadian Perspectives on Social Movements*, and *Gender Relations in Canada: Intersectionality and Beyond*.

Leslie Regan Shade is an Associate Professor in the Department of Communication Studies at Concordia University, Quebec, Canada. Her research focus since the mid-1990s has been on the social, policy, and ethical aspects of information and communication technologies (ICTs), with particular concerns towards issues of gender, youth, and political economy. She is the author of *Gender and Community in the Social Construction of the Internet* (Peter Lang, 2002), and co-editor of *Feminist Interventions in International Communication* (with Katharine Sarikakis, Rowman & Littlefield, 2008), two volumes in *Communications in the Public Interest* (edited with Marita Moll, Canadian Centre for Policy Alternatives), and with Moll *For Sale to the Highest Bidder: Telecom Policy in Canada* (CCPA, 2008) and the editor of *Mediascapes: New Patterns in Canadian Communication*, 3rd ed. (Nelson Canada). Articles have also appeared in *Continuum, The Gazette, Canadian Journal of Communication, Feminist Media Studies,* and *Government Information Quarterly*.

Mimi Sheller is Professor in the Department of Culture and Communication at Drexel University, Pennsylvania, USA. She is co-editor of the journal *Mobilities* and author of *Consuming the Caribbean* (Routledge, 2003); *Democracy After Slavery: Black Publics and Peasant Radicalism in Haiti and Jamaica* (Macmillan Caribbean, 2000); and is currently completing a book entitled *Citizenship from Below: Caribbean Agency and Modern Freedom* (for Duke University Press). She is co-editor with John Urry of *Mobile Technologies of the City* (Routledge, 2006), *Tourism Mobilities* (Routledge, 2004), and a special issue of *Environment and Planning A* on Materialities and Mobilities.

About the Contributors

Tamara Shepherd is a Ph.D. candidate in the Joint Doctorate in Communication at Concordia University in Montréal, Canada. She has published and presented papers on aspects of labour in user-generated content and social media, from a feminist political economy perspective. Her dissertation research looks at the implications of young people's cultural production on the Web for the development of new media policy.

Rob Shields is Henri Marshall Tory Research Chair and Professor in Sociology and in Art and Design at the University of Alberta. He is co-editor of the journal *Space & Culture* and author/editor of a dozen books including *Places on the Margin: Alternative Geographies of Modernity* (Routledge, 1991), *Lifestyle Shopping: The Subject of Consumption* (Routledge, 1992), *Cultures of Internet: Virtual Spaces, Real Histories, Living Bodies* (Sage, 1996), *Lefebvre: Love and Struggle* (Routledge, 1999), *The Virtual* (Routledge, 2003), and *What Is a City: Rethinking the Urban After Hurricane Katrina* (University of Georgia Press, 2008).

Phillip Vannini is Professor in the School of Communication & Culture at Royal Roads University in Victoria, BC, Canada. He is author/editor of eight books, including Body/Embodiment: Symbolic Interactionism and the Sociology of the Body (edited with Dennis Waskul, Ashgate, 2006), Understanding Society Through Popular Music (authored with Joseph Kotarba, Routledge, 2008), Authenticity in Culture, Self, and Society (edited with J. Patrick Williams, Ashgate, 2009), Material Culture and Technology in Everyday Life: Ethnographic Approaches (Peter Lang, 2009), The Cultures of Alternative Mobilities: Routes Less Travelled (Ashgate, 2009), Popularizing Research: Engaging New Media, Genres, and Audiences (Peter Lang, 2012), The Senses in Self, Society, and Culture: A Sociology of the Senses (authored with Dennis Waskul and Simon Gottschalk, Routledge, 2011), and Ferry Tales: An Ethnography of Mobility, Place, and Technoculture on Canada's West Coast (Routledge, 2012). He is also author of over sixty articles and book chapters, which have appeared or are forthcoming in such journals as Space & Culture, Journal of Transport Geography, Environment & Planning D, Journal of Contemporary Ethnography, Symbolic Interaction, The Senses and Society, Social Psychology Quarterly, Canadian Journal of Communication, Qualitative Inquiry and more. His research interests in mobility studies have concentrated on marine transportation. His ethnographic work on ferry passengers spans the fields of cultural studies, communication studies, cultural anthropology, cultural geography, and cultural sociology.

Index

ageism, 167
advertisements; airline, 29-30;
    in Chile, 245-246
    as cultural artifacts, 100;
    cruises, 29-30;
    mobile phone, 205-206, 212-213, 215;
    PAN AM, 108-109;
    rhetoric, 103;
    transparency, 206-208, 212-213;
    youth, 204, 215
Archipelagos, 23
automobile, 8, 79-4, 226-229;
    and cities, 80-2, 85, 90-4;
    and civic practice, 80-1;
    "auto-space", 83, 93;
    infrastructure, 83-4;
    space/planning see planning;
    and status, 259
    traffic see traffic;
    youth, 147

Brathwaite, Kamau, 37
Buses, 260-262

Canadian Radio-television and Telecommunications Commission (CRTC), 200-202
car *see* automobile
census, 46, 52-3
Certeau, Michel de, 159-160, 165, 171
Constellations of mobility, 199-201, 203-208, 238
cycling, 229-231

fares, 111, 261-262, 270-271
flight, 8-9, 100;
    commercial, 102-104, 108-110;
    historians, 101, 103, 113;
    international, 103-104, 108-112;
    PAN AM, 108-109;
    technology, 112-113

immobility, 239, 247-248, 251
Internet, 121;
    social activism, 122-126, 130-131, 134-135
infrastructure, 59-61;
    relation between humans, 73-5, 60-1
    Transantiago, 268-269

Metropolis project, 49
migration, 44-8; cities; 45-6, 50, 52, 80, 85;
    enclaves, 52; immigration, 50, 243;
    issues in research, 50; tracking, 52-5
mobility action chains, 144, 152;
    end of life, 151;
    late-life, 149-151;
    mid life, 147-149;
    newborns, 145-146;
    youth, 146-147
mobile solidarity, 131-133, 135-136

narratives, 61-2, 74-5;
    automobile, 84;
    chronology, 65-6;
    many voices, 71
New Social Movements, 129-130, 135

offshore economies, 32

Pan American Mobility Network, 4
place, 61-3, 188;

assembledge of, 63, 65, 74;
narratives of, 61-2;
rituals, 176
planning; automobile, 82-88, 90-94;
car oriented development
(COD), 91;
cities, 86-88;
cycling, 94;
machine roads, 85, 88-91, 94;
new urban roads, 85, 91-4;
park roads, 85-8, 93;
traffic see traffic;
and transportation, 263-266, 274
urban growth, 92
production of space, 79;
automobile see automobile
freeways, 82, 90;
planning, 82-4

roads, 79; machine roads, 85, 88-91, 94;
new urban roads, 85, 91-4;
park roads, 85-8, 93;

seawall, 59-65
Simmel, Gerog, 219-220
Slave Trade, 31-2
sociotechnical, 4-6
specialization, 47-8

technics, 3-4, 6
technicians, 2-4, 6, 16
technique, 3
technology, 237, 256; as extension of self,
141-143;
as ghosts, 188-191, 193;
haunting of, 190-191
traffic, 222-223, 231;
automobile, 223-224;
late-life, 149-150;
mobile looking, 220-222;
pedestrian, 223-225;
planning, 89-91
transnational mobilities, 237, 242, 250
travel writing, 28-9

Vieques Island, 35-6
virtual economy, 31-4
voice, 60, 71; bloggers, 67; NGO, 70;
objects, 60, 70-3;
publications, 68-9;
researchers, 71

walking, 224-226
Wennerlind, Carl, 31

Intersections in Communications and Culture

Global Approaches and Transdisciplinary Perspectives

General Editors: Cameron McCarthy & Angharad N. Valdivia

An Institute of Communications Research, University of Illinois Commemorative Series

This series aims to publish a range of new critical scholarship that seeks to engage and transcend the disciplinary isolationism and genre confinement that now characterizes so much of contemporary research in communication studies and related fields. The editors are particularly interested in manuscripts that address the broad intersections, movement, and hybrid trajectories that currently define the encounters between human groups in modern institutions and societies and the way these dynamic intersections are coded and represented in contemporary popular cultural forms and in the organization of knowledge. Works that emphasize methodological nuance, texture and dialogue across traditions and disciplines (communications, feminist studies, area and ethnic studies, arts, humanities, sciences, education, philosophy, etc.) and that engage the dynamics of variation, diversity and discontinuity in the local and international settings are strongly encouraged.

LIST OF TOPICS

- Multidisciplinary Media Studies
- Cultural Studies
- Gender, Race, & Class
- Postcolonialism
- Globalization
- Diaspora Studies
- Border Studies
- Popular Culture
- Art & Representation
- Body Politics
- Governing Practices
- Histories of the Present
- Health (Policy) Studies
- Space and Identity
- (Im)migration
- Global Ethnographies
- Public Intellectuals
- World Music
- Virtual Identity Studies
- Queer Theory
- Critical Multiculturalism

Manuscripts should be sent to:

Cameron McCarthy OR Angharad N. Valdivia
Institute of Communications Research
University of Illinois at Urbana-Champaign
222B Armory Bldg., 555 E. Armory Avenue
Champaign, IL 61820

To order other books in this series, please contact our Customer Service Department:
(800) 770-LANG (within the U.S.)
(212) 647-7706 (outside the U.S.)
(212) 647-7707 FAX

Or browse online by series:
www.peterlang.com